PHOTOVOLTAIC WATER PUMPING SYSTEMS

PHOTOVOLTAIC WATER PUMPING SYSTEMS

Concept, Design, and Methods of Optimization

TAMER KHATIB
Energy Engineering and Environment, An-Najah National University, Nablus, Palestine

DHIAA HALBOT MUHSEN
Computer Engineering Department, Mustansiriyah University, Baghdad, Iraq

ELSEVIER

ACADEMIC PRESS
An imprint of Elsevier

Academic Press is an imprint of Elsevier
125 London Wall, London EC2Y 5AS, United Kingdom
525 B Street, Suite 1650, San Diego, CA 92101, United States
50 Hampshire Street, 5th Floor, Cambridge, MA 02139, United States
The Boulevard, Langford Lane, Kidlington, Oxford OX5 1GB, United Kingdom

Notices
Knowledge and best practice in this field are constantly changing. As new research and experience broaden our understanding, changes in research methods, professional practices, or medical treatment may become necessary.

Practitioners and researchers must always rely on their own experience and knowledge in evaluating and using any information, methods, compounds, or experiments described herein. In using such information or methods they should be mindful of their own safety and the safety of others, including parties for whom they have a professional responsibility.

To the fullest extent of the law, neither the Publisher nor the authors, contributors, or editors, assume any liability for any injury and/or damage to persons or property as a matter of products liability, negligence or otherwise, or from any use or operation of any methods, products, instructions, or ideas contained in the material herein.

Library of Congress Cataloging-in-Publication Data
A catalog record for this book is available from the Library of Congress

British Library Cataloguing-in-Publication Data
A catalogue record for this book is available from the British Library

ISBN 978-0-12-821231-8

For information on all Academic Press publications
visit our website at https://www.elsevier.com/books-and-journals

Publisher: Brian Romer
Acquisitions Editor: Lisa Reading
Editorial Project Manager: Alice Grant
Production Project Manager: Poulouse Joseph
Cover Designer: Christina Bilbow

Typeset by SPi Global, India

To Zain and Sara, just to be a fair father,
Tamer Khatib

To those who gave me the true meaning of life, my dear wife Alaa,
my daughter Manar, and my son Ahmad,
Dhiaa Halboot Muhsen

Contents

About the authors

Associate Professor Dr. *habil* Tamer Khatib is a photovoltaic power systems professional. He holds a B.Sc. degree in electrical power systems from An-Najah National University, Palestine, as well as an M.Sc. degree and a Ph.D. degree in photovoltaic power systems from the National University of Malaysia, Malaysia. In addition, he holds Habilitation (the highest academic degree in German-speaking countries) in renewable and sustainable energy from Alpen Adria Universitat, Klagenfurt, Austria.

Currently, he is an associate professor of renewable energy at An-Najah National University and the director of the An-Najah Company for Consultancy and Technical Studies. He is also the chair of the IEEE Palestine subsection.

So far, he has 2 patents, 3 books, and 122 research articles while his current h-index is 32. He has supervised 4 Ph.D. studies, 16 master studies, and 50 bachelor studies.

He is a senior member of IEEE, the IEEE Power and Energy Society, the International Solar Energy Society, and the Jordanian Engineers Association.

His research interests mainly fall in the scope of photovoltaic systems and solar energy fundamentals. These interests include PV system design and optimization, modeling and control of PV systems, hybrid PV/wind systems, hybrid PV/diesel systems, grid-connected PV systems, sun trackers, MPPT technology, inverters in PV systems, solar chargers, batteries and charge controllers, solar energy fundamentals, solar energy prediction, AI applications for solar energy and PV systems, wind power systems, wind chargers, and wind energy modeling and prediction.

Assistant Professor Dr. Dhiaa Halbot Muhsen earned a B.Sc. degree in electrical engineering from the University of Baghdad, Iraq, in 2001, as well as an M.Sc. degrees in computer and control engineering from Baghdad University, Iraq, in 2004. In addition, he holds a Ph.D. degree in computer engineering from the University of Tenaga National, Malaysia. Currently, he is an assistant professor in the computer engineering department at Mustansiriyah University. He earned his Ph.D. under the supervision of associate professor Dr. *habil* Tamer Khatib.

So far, he has 21 research articles while his current h-index is 9.

His current research interests include modeling of photovoltaics, modeling of PVPS, sizing of PVPS, load scheduling in smart grids, evolutionary algorithms (EA), and multicriteria decision-making (MCDM) algorithms.

Preface

Solar energy is considered a renewable and environmentally friendly energy source. Meanwhile, photovoltaic technology (PV) is the most popular technology that directly utilizes solar energy and converts it to a direct current. One of the most popular applications of PV power systems is the PV water pumping system (PVPS). The PVPS utilizes the power generated from the PV module to power a pumping system for different water pumping applications such as providing drinking water for domestic use and irrigation. A properly designed PVPS can be efficient and economically competitive to grid-connected or diesel generator (DG)-based pumping systems, especially in rural areas. However, the low energy-conversion efficiency of PV modules, the nonlinearity of the PV module/array I-V characteristics, and the unique maximum power operation point are still the major challenges of this technology.

This book aims to provide a concept of a photovoltaic water pumping system, that system's components, the mathematical models, and the methods of design, including novel artificial intelligence methods. Thus, this book includes six chapters. The first chapter proposes a general introduction while the second chapter reviews the photovoltaic water pumping concept, performance, reliability, control strategies, and sizing methods. The third chapter discusses the mathematical models of system components. The fourth and fifth chapters propose design methods, including the intuitive methods, the numerical methods, and the artificial intelligence methods. Examples are provided with open source MATLAB codes. The last chapter discusses the feasibility and sensitivity of the photovoltaic water pumping system, considering different types of storage and backup sources.

Acknowledgments

The authors would like to thank Professor Dr. Azah Mohamed, Assistant Professor Dr. Haider Tarish Haider, Assistant Professor Dr. Issa Ahmed Abed, and Eng. Alia Saleh. Moreover, the authors would like to extend their thanks to Elsevier's team for their cooperation with special thanks to Lisa Reading, Danielle McLean, Alex Ford, Poulouse Joseph and Ashwathi Aravindakshan.

CHAPTER 1
Introduction

Contents

1.1 The need for water pumping using photovoltaic systems

Recently, water demand has increased due to the increasing population, and the availability of water has become more crucial than ever before. A source of energy to pump water is also a big problem in many developing countries. Meanwhile, developing a grid system is often too expensive, especially in rural villages that are located too far away from existing grid lines. On the other hand, dependency on an imported fuel is not a reliable option due to political issues, foreign exchange rate fluctuation, and the economies of many developing countries, which are the supply source. Even if fuel is available within the country, transporting that fuel to remote, rural villages can be difficult.

The use of renewable energy is attractive for water pumping applications in rural areas of many developing countries. The transportation of renewable energy systems, such as photovoltaic (PV) pumps, is much easier than other types because they can be transported in pieces and reassembled on site.

A solar pumping system is a technique that has been used recently to provide a water supply from water sources such as artificial ponds, water wells, boreholes, or rivers. In places where there is no power grid, solar pumping is often the most obvious solution, as the installation and operation of renewable energy systems is easier than other systems. This is because they can be assembled in pieces onsite and do not imply running fuel that needs to be transported. A photovoltaic (PV) pumping system can help agriculture fields become more sustainable, as these systems can minimize the overirrigation of the field. In areas where water is underground, the most common way to

Photovoltaic Water Pumping Systems
https://doi.org/10.1016/B978-0-12-821231-8.00001-9

lift water is hand pumps with primitive tools such as a Shadoof (Arabic name of a manual and conventional pumping tool). The other option is diesel generators, which are much better than the Shadoof. However, diesel generators have negative impacts on health and the environment as well as influences on irrigation because of the potential leakage. Moreover, due to the high maintenance of a diesel generator and fuel transportation costs, the cost of the pumped water is high as \$1 per m^3.

The advantages of PV-powered pumps are low maintenance, no pollution, easy installation, reliability, possibility of unattended operation, and the capability to be matched to demand. Meanwhile, the disadvantages are the high initial cost and variable water production. However, if the designer is familiar with a well's specification, the needed storage system, the natural terrain surrounding the well, and the manufacturers' data on available pumps and other equipment, a reliable pumping system can be provided.

1.2 Types of water pumping systems

Water pumping systems are classified depending on the type of pump used (AC or DC) and the configuration of the systems. There are two configurations of water pumping systems: standalone systems and hybrid systems. Hybrid systems in general use two energy sources such as a wind turbine, diesel generator, or PV. The AC standalone water pumping systems in general consist of a PV array, an inverter with a centralized maximum power point tracker, and a pump. Such a system can have a block of batteries as a backup power supply, but because the system cost is a very important factor, systems are designed in a way to meet the village demand during the solar day without any need for batteries.

There are two broad categories of pumps used in standalone PV systems around the world: rotating and positive displacement. There are many variations on the designs of these two basic types. Examples of the rotating pump type are centrifugal, rotating vane, or screw drive. These pumps move water continuously when power is presented to the pump. The output of these pumps is dependent on head, solar radiation (current produced), and operating voltage. They are well suited for pumping from shallow reservoirs or cisterns. They can be tied directly to the PV array output, but their performance will be improved by using an electronic controller such as a linear current booster to improve the match between the pump and the PV array. Positive displacement pumps move "packets of water." Examples are diaphragm pumps and piston pumps (jack pumps). These pumps are typically used for pumping water from deep wells. Their output is nearly independent of the pumping head and proportional to solar radiation. Jack pumps should not be connected directly to a PV array output because of the large load current changes during each pump cycle.

In general, peak power controllers are recommended. The controllers adjust the operating point of the PV array to provide maximum current for motor starting and then keep the array operating at the maximum power conditions. In some cases, designers use batteries between the jack pump and the array to provide a stable voltage source to start and operate the pump.

Usually they are not sized to provide night-time pumping, but only to give stable system operation. Pumps are also categorized as surface or submersible. Surface pumps have the obvious advantage of being more accessible for maintenance. When specifying a surface pump, you must distinguish between suction and lift. A pump may be installed a few feet above the water level, with a pipe from the pump to the water. The maximum length of the pipe is determined by the suction capability of the pump. The pump may then "lift" the water to a storage tank above the pump. The elevation of the storage tank is determined by the lift capability of the pump. Most submersible pumps have high lift capability. They are sensitive to dirt sand in the water and should not be run if the water level drops below the pump. The type of pump depends on the water required, the total dynamic head, and the capability of the water source.

Meanwhile, both rotating and displacement pumps can be driven by AC and DC motors. The choice of motor depends on the water volume needed, the efficiency, the price, the reliability, and the availability of support. DC motors are an attractive option because of their compatibility with the power source and because their efficiency is usually higher than that of AC motors. However, their initial cost is higher, the selection may be limited in some countries, and the brush-type motor requires periodic maintenance. Some brushless DC motors are available and promise improved reliability and decreased maintenance. AC motors require a DC-to-AC inverter, but their lower price and wider availability are advantages.

In water pumping systems, storage can be achieved by using batteries or by storing the water in tanks. Adding batteries to a system increases the cost and decreases reliability. Water storage is better for most applications. However, considerable evaporation losses can occur if the water is stored in open tanks or reservoirs. Meanwhile, closed tanks large enough to store several days of water supply can be expensive. In some countries, these tanks are not available or the equipment necessary to handle, move, and install the tanks may not be available. Also, any water storage is susceptible to vandalism and pollution.

1.3 Irrigation systems

The distribution system in agriculture applications consists of the conveyance network that delivers water to irrigate the field. The style of this network depends on the adopted irrigation field technique.

Using an effective irrigation method is vital to reduce the losses of infiltration of the water, especially as the unit cost depends directly on the output water required. The following are the common methods for irrigation techniques:

- Flood irrigation: the water in this technique is distributed throughout the field. Thus, it consumes large quantities, and the water demand is irregular. Therefore, the pump size is determined by the peak demand at flooding time. Anyway, such a technique is not recommended for PV pumping applications.
- Channel irrigation: water is filled in furrows. With such a practice, there is high loss due to infiltration and run-off of the water surface. This technique is also not recommended for PV pumping.
- Sprinkler irrigation: despite its high efficiency, it is not advisable for PV pumping systems because it needs high static heads to sprinkle the water.
- Basin and hose irrigation: the basin is an area around the tree with which to fill the water. It is surrounded by bunds to block the flowing of the water to the adjacent areas. The water is delivered to the basins using a hose. It's the most effective of the traditional irrigation techniques because of its simplicity and high efficiency. However, it's efficient if the labor cost is low.
- Drip/trickle irrigation: the water is delivered through the main pipe and then by smaller pipes. It has high efficiency because it has large pipes with low flow rates.

  This technique is favorable for PV pumping applications. Table 1.1 summarizes the efficiencies and the required head of the irrigation systems:

Table 1.1 Efficiencies and typical head of irrigation systems.

Irrigation technique	Rate efficiency (%)	Required head (m)
Flood	40–50	0.5
Channels/furrow	40–60	0.5–1
Sprinkle	70	10–20
Basin/hose	90	1–2
Drip/trickle	85	1–2

CHAPTER 2

Photovoltaic water pumping systems concept

Contents

2.1 Photovoltaic water pumping system theory

A PV pumping system consists of five major components, which are PV array, a power conditioning unit, pump-motor load, water tank storage, and pipe distribution system. PV pumping systems can be battery-coupled or directly coupled. The battery-coupled type has PV panels, charge regulator, batteries, pump controller, tank, and DC pump. On the other hand, the directly coupled type has no batteries, and thus water has to be stored in the tank so as to be used at night or on cloudy days. A PV pumping system has two configurations based on the water source. Such a system may be either surface water such as a river or artificial pond or subsurface such as wells.

2.1.1 Motor-pump set in a PV pumping system

Motors are categorized into alternating current (AC) and direct current (DC), where each category has its own types and applications that are matched with the motors' characteristics. The DC permanent magnet (brushed and brushless) and the squirrel cage induction motors are the

Photovoltaic Water Pumping Systems
https://doi.org/10.1016/B978-0-12-821231-8.00002-0

common types used in PV pumping applications. Selecting the required type of motors relies on the size requirement as well as the type of water source. The motor properties of efficiency, availability, and price are usually considered in selecting the motor. In general, the power demand is the dominant parameter in choosing the appropriate motor. For instance, for 3 HP of mechanical load, DC permanent magnet motors are used. Meanwhile, for 3–10 HP, DC wound field motors are used. On the other hand, AC induction motors are used for high power demand that is above 10 HP. Generally, DC motors have higher efficiency than AC motors. Moreover, they don't need an inverter as PV array generates DC power. However, DC motors need periodic replacement due to the mechanical moving parts (commutator), usually after 2000–4000 h. By contrast, AC motors are used for high power demand and they are cost-effective. Fig. 2.1 shows the types of motors used in photovoltaic water pumping systems.

Pumps can be classified into rotodynamic (centrifugal) and volumetric (positive displacement). Each category works at different pumping conditions based on its characteristics. Centrifugal pumps mainly depend on the head, input current, and operating voltage. The pumping is caused due to the pressure difference created between the inlet and the outlet of the pump. The pump's water output increases proportionally with the speed of rotation. They have an optimum efficiency at a certain design head and rotation speed. Centrifugal pumps can be connected directly with PV modules. However, power regulation devices can be connected to enhance their performance such as linear current booster. On the other hand, positive displacement pump's flow rate of water does not depend on the head, but on the speed proportionally. Unlike centrifugal pumps, the match between the

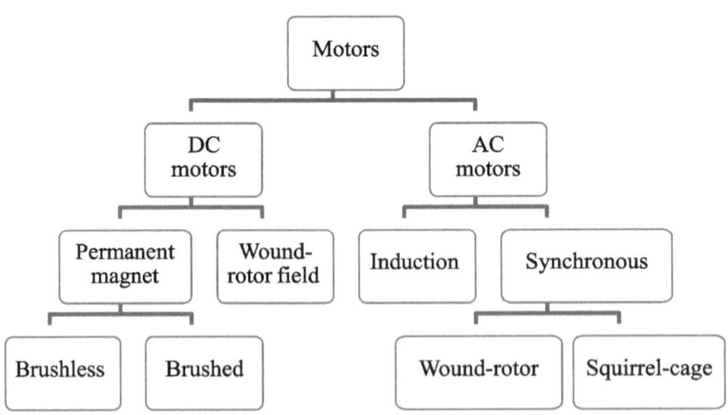

Fig. 2.1 Motor classifications chart.

positive displacement pumps and the PV array is not suitable. This is because at the beginning of the operation, the motor that is running the pump needs high torque, which means a high starting current.

After all, selecting the proper pump depends on multiple parameters such as the water requirement per day, the source of water, and the pumping head. In general, volumetric pumps are suitable for water flow under $15\,m^3/day$ and pumping head between 30 and 150 m. Meanwhile, submersible centrifugal pumps work well for high flow rate ($25–100\,m^3/day$) and medium pumping heads (10–30 m).

2.1.2 Static and dynamic heads of water pumping system

Total dynamic head (TDH) is one of the important parameters in designing a pumping system because pumps are specified based on a particular TDH and flow rate. TDH is divided into a static head and a dynamic head. The static head is the distance between the water surface level of the well to the top of the storage tank (if any). The dynamic head is also divided into a friction head and the drawdown level of the well. The drawdown is the drop of water in the well due to the pumping. Meanwhile, friction loss is caused by the length of the pipes, valves, and other fittings. Friction loss depends on the inside diameter of the pipe, the roughness of the pipe, and the type of flow. Because friction loss depends on the roughness of the pipe materials, selecting the appropriate material is vital. Many factors are taken into consideration in selecting pipes such as pipe roughness, pipe's strength and durability, soil type, and cost. Many materials are used in the piping system such as galvanized iron and plastic pipes. Plastic pipes have many commercial types such as PVC, C-PVC, and PE.

2.1.3 Storage system for photovoltaic water pumping system

Storage in PV pumping system means either energy storage or water storage. It is favorable to store water rather than energy due to some reasons such as batteries reduce system's overall efficiency. Moreover, batteries are expensive and need maintenance, which adds more cost to the system. Therefore, water storage is more economical and simpler. The proper selection of a storage tank is important to give the suitable pressure and volume for the used irrigation technique. Moreover, it is also important to design it in an optimal way so as to deliver water for crops without any interruption. It is also vital as the reliability of the PV system is lower than conventional systems. In addition, plants are irrigated during periods when the solar intensity is low (early morning or late afternoon) as the evapotranspiration is low.

Typically, the storage tank is designed to withstand 3 days without any solar radiation to meet the required water demand. The storage tank comes with different configurations. Basically, there are ground, elevated, and hydropneumatic reservoirs. Based on the shape, the elevated reservoir can be classified as a rectangular, an intze, or a cylindrical tank. A cylindrical tank, for example, has features of simplicity in geometry. It is also found that a cylindrical tank is very economical for lower volume capacities.

Regarding the material, there are common materials that the tanks can be made of such as stainless steel, concrete, carbon steel, fiberglass, and plastic. Each material has its own specifications of price, weight, temperature, pressure handling, and corrosion resistance. There are some factors that should be taken into consideration in designing water storage tanks for PV pumping system, such as:

- The tank size should be bigger than the tanks that are used for conventional systems because of the uncertainty of solar radiation. The type of application is important as well. For example, the tank of a PV pumping system for agriculture is usually larger than the tanks used for water supply and livestock.
- The size of the distribution system.
- Available water resources: in some cases, storage tanks are not needed, especially where there is surface water with a year-round flow.

2.2 Installation of photovoltaic water pumping system

Many failures of PV pumping systems are caused by pump problems. The PV power supply has much higher reliability than the pump/motor subsystem. A good installation of the pumping hardware will increase reliability. Some things to watch for are described below:

Varying water levels: The water level in a well may vary seasonally, daily, or even hourly. The water level in some wells in rocky areas has been reported to drop as much as 75 ft during pumping. The pump must be mounted to keep the water inlet below the water level at all times. If the replenishment rate of a well is lower than the maximum possible pumping rate, a level switch or mechanical valve should be included to protect the pump from operating dry. Float switches should be used on storage tanks if the volume of the tank is smaller than the daily pump rate. This will prevent wasting water or pump damage due to overheating.

Protect the pump input: Sand is a primary cause of pump failure. If the well is located where dirt and sand may be pulled into the pump, a sand screen should be used. Most pump manufacturers offer this option or they can recommend methods to limit the risk.

Ground the equipment: Water pumps attract lightning because of the excellent ground they provide. If possible, do not locate the pump system on high ground. Consider erecting lightning rods on higher terrain around the pump. Ground the pump motor, the array frame, all equipment boxes, and one system conductor to the well casing (if metal) or to a bare conductor running down to the water level. Never use the pipe string to the pump as a ground because the ground would be interrupted when maintenance is being performed. The use of movistors to protect electronics is recommended in areas prone to lightning.

Avoid long pipe runs: when long pipe runs are used in a well with a jack pump, losses may significantly increase. Friction losses depend on pump maintenance. The size of the pipe, the length, the ameter should be larger than the flow rate, and the number of bends pump cylinder. As the output of PV system is limited and varies throughout the day, it is particularly important to keep friction losses low. Pump system efficiency can drop to near zero if a large friction loss occurs. Thus, it is important to limit the friction loss to less than 10% of the head. This can be done by oversizing the pipe, eliminating bends and junctions, and reducing flow rate. Data on pipe size and friction rates are available from pump manufacturers. Pump leathers to be changed by pulling the sucker rod without pulling the pipe.

Protect the control equipment: All electronic control equipment should be housed in weather-resistant boxes. All wires should be approved for outdoor use or installed in a conduit. Any cables used for submersible pumps should be appropriate for that application. Pump manufacturers will give recommended wire types for their equipment.

Use steel pipe: Steel pipe is recommended for use in the well, particularly if submersible pumps are used. Plastic pipe may break. However, plastic pipe provides an inexpensive way to run water from the well to the storage tank or end user.

Protect the well: Use sanitary well seals for all wells. Bury pipes from the wellhead to the tank at a depth that will insure that the pipe won't be broken by traffic or during future trenching or excavation. Mark pipe runs for future reference.

Design consideration of water tanks: Some of the most important factors to consider in designing water tanks follow:

- *The power source type*: The size of the water tank should be greater for PV water pumping systems than for conventional systems because of the intermittent nature of the power source. Poor solar radiation or windless days creates a significant problem in fulfilling the daily water demand. The size of the water tanks for conventional systems depends only on the peak and average daily water demands. However, in the case of PV and wind pumps, the size of the water tanks depends on those factors and local weather conditions.
- *The geographical location of the system*: Geographical location is very important for a PV pumping system. Sunny areas make PV pumps much cheaper and their use ideal in certain locations. Similarly, the size of the water tanks can be smaller in such locations. On the other hand, the size of the water tanks should be much larger for areas of low solar radiation or low wind conditions.
- *The type of end user*: Unlike those in urban areas, PV pumps installed in rural communities are often designed based on minimum water demand requirements.
- *The type of application*: Water reservoirs designed for irrigation purposes should be very large compared to those for other water supply purposes. Water reservoirs designed for cattle watering and community use typically depend on per capita water consumption. As a rule of thumb, there should be 3 days of storage for a community water supply and 5 days of storage for cattle watering.
- *The size of the water distribution networks*: Because they can hold a huge amount of water, large water distribution systems also act as water reservoirs. Small distribution networks with water service at only a few points do not have such options. Many of these services are in rural villages where distribution pipes are usually small. In such cases, it is essential to pump water to holding tanks and then on to distribution networks.
- *Availability of other water resources*: Large tanks may be unnecessary if there is a great amount of annual rainfall in the area. Rainwater can be collected and used for cattle watering and for washing clothes in rural areas. Surface water that flows year round (such as a river) can also provide such services to reduce the need for large-capacity water tanks. Unlike with conventional pumping systems, estimating the proper size of a PV pumping system, including the water tank, requires a detailed evaluation of each component and the system as a whole. As water tanks often cost less than a PV array, installing a larger water tank (instead of a larger PV pumping system) allows for the possibility of storing water for low

solar radiation. Every means to reduce costs must be considered when designing PV pumping systems. For example, for PV pumping systems, using seasonally adjustable tilt angles can increase the daily water production by 5–10%. It is also necessary to supply enough water for days that feature poor solar radiation, and to use the surplus water produced on days with optimum sun. Consecutive days with poor solar radiation are especially difficult when trying to meet daily water demand, but can be offset with a surplus.

Water tanks for conventional supply systems are designed to hold 30%–40% of the total daily water demand. The bare minimum tank size for any community can be estimated by simulating typical 1 year daily weather data and estimating the water balance from the daily water demand and production over the year. The water balance is the difference between the daily water production and water demand. A negative water balance demonstrates the extra amount of water required over the day. The worst consecutive days of poor solar radiation conditions are reflected on the values of the water balance. The extra amount of water required for these consecutively poor days is calculated by adding the daily water balance of those days with the negative sign that is equal to the amount of extra stored water required to meet the water demand. Because there may be several consecutive poor days, those days with the largest water requirement are exactly the minimum size of the water tank required to fulfill the community's water demand. In the case of hybrid systems, water tanks can be sized like conventional systems because there is a good possibility of getting enough energy to pump water either from wind, sun, or a standby diesel system.

2.3 Photovoltaic water pumping system field performance

As a fact, the performance of any PVPS depends on many factors such as meteorological variables (solar radiation, ambient temperature, wind speed, humidity, and shadow effects), PV module specifications (conversion efficiency and tilt angle), and motor pump–hydraulic system characteristics (I-V characteristic of the motor pump set as well as static, dynamic, and friction loss heads). In this section, PVPS system reliability, feasibility, and field performance are reviewed based on a number of system performances reported in the literature. Eventually, a summary of these results is provided. The latest research articles are reviewed and system performance is reported.

In general, the annual productivity rate (volume of pumping water per 1 kWp) needs to be calculated in order to assess the feasibility of a PVPS.

Also, the PV array, subsystem, and overall system efficiencies, including the performance factor (the ratio of field performance to the theoretical performance), can be used to check the feasibility of a PVPS. In addition, the yield factor (the ratio of the average daily pumped water to the rated pumped water of PVPS) and the capacity factor (the ratio of actual annual pumped water to the annual pumped water of a PVPS that operates at full rating) are other factors that can be utilized to evaluate the productivity of a PVPS. Moreover, the cost of a water pumping unit is an important assessment factor. In [1], the performance of a PV array in a PVPS is evaluated. According to the authors, the field conversion efficiency of the PV array reaches 8.5%. Furthermore, Bucher et al. [2] conducted a feasibility study of a PVPS based on a comparison between PVPS and a DG-based water pumping system based on experimental results. According to the results, a large PV water pumping system is required as compared to the DG-based system, which consequently results in a high initial investment. The efficiency of the available pumping system (motor/centrifugal pump assemblies) is in the range of 35%–65%. Meanwhile, the overall system performance efficiency is in the range of 3%–10%. The unit cost is found to be about 0.60 USD/m^3, which is less than the DG-based pumping system (1 USD/m^3). Posorski et al. [3] evaluated 90 PV-water pumping systems installed in seven different countries (Argentina, Jordan, the Philippines, Tunisia, Zimbabwe, Indonesia, and Brazil). The average overall efficiencies of the systems was 3%. The cost of water discharge is 1 USD/m^4 and 0.9 USD/m^4 for PVPS and DG-based water pumping systems, respectively. In [4], the steady-state performance of a PVPS is analyzed. The adopted system consists of a 6.048 kWp PV array, a storage battery with a capacity of 1060 Ah, and a centrifugal pump. The switched reluctance motor that is used in the pumping system has an overall efficiency above 85%. According to the results, most of the losses are pumping losses (31.82%) and therefore the efficiency of the pumping system can be improved by selecting an efficient pump. The matching efficiency of the proposed system and the PV array reaches 95%.

In Ref. [5], the performance of a number of PVPSs in four distinct sites in Algeria (Tamanrasset, Oran, Algiers, and Bechar) was analyzed. These PVPSs have different hourly load demands, PV module types, PV array configurations, storage tank capacities, pumping heads, and pump types. The influence of load profiles is illustrated by a factor called relative deficit (the ratio of the total demand deficit to the total of the demanded water). The results show that there is a linear relationship between the relative deficit and the pumping head. In addition, the relative deficit value depends on the storage tank of the system. The PVPS with a multistage pump performed the best for a high pumping head, whereas the motor-pump and total efficiencies are 23% and 1.5%, respectively. In [6], the performance of a PVPS in Jordan is investigated for 13 distant underground water wells. The studied wells have water depths in the range of 18–75 m. Meanwhile, the system water demands are in the range of 36–130 m^3/day. The pumping power is in the range of 0.5–1.94 kW. Hammad [6] proposed a new factor called the pumping factor to compute the monthly performance.

In Ref. [7], four PV water pumping systems installed at four locations in Kairouan in the Tunisia region were evaluated. Each system consists of a PV array with a capacity of 2.1 kW that supplies a three-phase 1.5 kW squirrel cage motor via a 3 kVA three-phase DC/AC inverter. The motor is coupled with an NE 42-30 Pleuger multistage submersible centrifugal pump. A water storage tank capacity of 40 m^3 volume is utilized in this system. The four systems deliver daily water in the range of 6.5–30 m^3/day with a head from 65 to 112 m. The maximum subsystem efficiency is 30% at a 65 m pumping head, 2.12 m^3/h flow rate, and 2840 rpm as the rated speed of the motor. According to the results, the optimal water flow rate is 2.7 m^3/h with an overall system efficiency of 2.5. In Ref. [8], the performance of three PVPSs with different PV array configurations was studied. The performance was investigated based on experimental results to choose a feasible pumping system to provide 4–6 m^3 per day with a pumping head in the range of 19–35 m. The first system (PVPS1) consists of a PV array that has two parallel strings with one module per string. Meanwhile, the second system (PVPS2) contains a PV array with a different configuration (two modules connected in series). The third system (PVPS3) consists of a PV array that has two parallel strings and two PV modules in series in each string. The systems are tested experimentally for water head in the range of 10–35 m. The PV array is operated with an efficiency of 12%. The results show that the production of PVPS1 is more than PVPS2 for a pumping head that is less than 30 m. Meanwhile, PVPS3 is the best with the aforementioned pumping head values. Furthermore, PVPS1 has the best overall system efficiency (3% at a 25 m head) while for a pumping head that is more than 25 m, PVPS2 is better in terms of overall efficiency (3.8% at 35 m head). Similarly, in Ref. [9], a comparative study of a PVPS with a different configuration is conducted. Four systems with different PV array configurations are adopted in this study. The first configuration (PVPS1) consists of three parallel strings, and each string comprises six 75 W modules that are connected in series (6S × 3P). The second (PVPS2), third (PVPS3), and fourth

(PVPS4) system configurations are 12S × 2P, 8S × 3P, and 6S × 4P, respectively. These systems are tested experimentally for a sunny solar day and based on an 80 m water head. According to the results, PVPS3 supplied a maximum daily water volume of 22 m^3 and had the maximum average pumping efficiency (45.06%) as compared to other configurations. Also, the overall efficiency of the fourth system was more constant during the operation period, except in low sunlight periods. On the other hand, the electric power provided by PVPS3 is the most suitable for the DC pump so as to assure maximum power operation.

In Ref. [10], the performance of a PVPS under different water demand profiles and two static water head values was investigated. The system consisted of a 1.5 kW PV array, a 1 horsepower permanent magnet DC motor, and a centrifugal pump with a flow rate and input power of 80 L/min and 750 W, respectively. The PV array comprises two parallel strings, and each string includes 15 modules connected in series. A 51 Wp PV module is used in this work. According to the results, the pumping water rate is in the range of 6–65 L/min with 6–8 pumping hours. The motor pump efficiency is in the range of 12%–30%. In [11], an economic study between two PVPSs powered by diesel and photovoltaic in Jordan was done. The daily water demand of the station was 45 m^3 at a 105 m pumping head. The systems consist of a 5.9 kWp, a 7.5 kVA three-phase DC/AC, and a 5.5 kW submersible motor pump with a flow rate of 8 m^3/h. Two storage tanks with 55 m^3 storage capacity are used for each system. According to the results, the PVPS is more cost-effective as compared to the DG-based system. In [12], the authors discuss the performance, reliability, and efficiency of different PVPSs that use two types of pumps (helical and diaphragm pumps). According to the experimental results, the maximum pump efficiency is 48% and 60% for the diaphragm and helical pumps, respectively. The overall system efficiencies for the PVPSs that use the diaphragm and helical pumps are 5% and 7%, respectively. In [13], the authors analyzed the performance of a PVPS to determine the overall system efficiency and the water pumping volume. The PV array used in the system is configured as six parallel strings, and each string contains two 51 Wp modules that are connected in series. A 400 Ah lead acid battery, a 1 kW DC charge controller, a 100 W submersible pump, and a water reservoir of 0.5 m^3 capacity are utilized. Two phases of operation are applied to this system. In the first phase, the PV array charges the batteries without powering the pump. In the second phase, the PV array is disconnected and the battery bank powers the pump. The results showed that the overall system efficiency is 5% with a 30 m total head. In [14], the feasibility of PVPSs in Turkey is discussed. Two types of PVPSs are used in this study. According to the results, the water unit costs of PVPS1 and PVPS2 are 0.24 US$/$m^3$ and 0.26 US$/m3, respectively.

In Ref. [15], the performance of a PVPS located in India is studied. The system is tested experimentally with a pumping head of 5 m for nine operating hours. The experimental results show that the system has an average daily production of 38 m^3 with a flow rate in the range of 0.02–0.14 m^3/min. According to the results, the overall system efficiency is 2.75%. In Ref. [16], a PVPS is proposed for a spesific location in Indonesia. The system aims to pump water for a 1400 m horizontal distance with a 218.34 m head. According to the results, the system starts pumping water at a solar radiation level of 300 W/m^2 while the flow rate was in the range of 1.44–3.24 m^3/h. In [17], the influence of meteorological variables (solar radiation and temperature) on the performance of a 2.1 kWp PVPS in Tunisia is studied. The system is equipped with a 3 kVA/127 VDC/AC inverter, a 1.5 kW 127 V squirrel cage motor that is driving a centrifugal pump, and a storage tank. The system is modeled and simulated using MATLAB/Simulink. The simulation results show that the overall efficiency of the system is about 3%. Meanwhile, the average daily production is found to be 7.7 m^3/day in January and 14.7 m^3/day in July. In [18], the influence of the pumping head on the performance of a PVPS in Saudi Arabia is studied. The system consists of a 1.8 kW photovoltaic array, a SQF5-2 submersible helical pump, and a storage tank. The system is tested with four different heads (50, 60, 70, and 80 m). The results illustrate the dependency of the water flow rate and total system efficiency on the solar radiation and pumping head. According to the results, the system with a 50 m pumping head has the best total system efficiency (about 7%) at low solar radiation times. With the 80 m water head, the highest total system efficiency (around 6.6%) is recorded at medium and high solar radiation only. In addition, the best average efficiency along the operation hours is 4.4% with a pumping head of 80 m. In Ref. [19], the feasibility of using PVPS in Egypt is studied. According to the results, the cost of the water pumping unit by PVPS is cheaper than the unit pumped by a DG-based system. However, the authors concluded that the cost of a water pumping unit is more sensitive to the price of PV modules than the PV array's lifetime. In Ref. [20], the influence of a shaded photovoltaic array on the performance of PVPS in Algeria is studied. Ten different partial shading scenarios are applied to make explicit the effect of shadow on the PVPS performance. According to the results, 50% of the PV production is reduced due to the PV array shadow where the overall system efficiency is degraded from 1.299% (without shading) to 0.83% (with shading). In addition, two shadow cases led to a system shortage. Finally, in Ref. [21], the authors investigated the feasibility of replacing the fossil-fuel source with a renewable energy source to power a water pumping system in Nigeria. The results show that solar and wind sources are more cost-effective than a DG-based water pumping system. Table 2.1 shows a summary of all the previously reviewed works.

2.4 Photovoltaic water pumping system design procedures

In fact, the high initial cost of the PV array is one of the important drawbacks of PVPS. Therefore, many researchers are focusing on the optimal size of a PV array as well as other components such as the storage unit and inverter to meet the required load at minimum cost [22–25]. In general, there are three main methods for sizing PV systems: intuitive, numerical, and analytical [26, 27]. The simplest one is the intuitive method, which is based on the worst month or the average monthly solar radiation in sizing the PV array and storage units. According to [28], using this method may lead to an over- or undersizing of the PV system, which consequently increases the system cost or decreases the reliability of the system. Thus, the intuitive method is only convenient to be used for estimating initial and rough approximation sizes of a PV system [29]. On the other hand, in the numerical method, hourly meteorological data are used to simulate the system so as to describe its hourly energy flow. After that, based on system performance, possible system configurations that achieve a specific level of reliability are found. Then, an evaluation based on specific constraints is applied to these configuration so as to pick the suitable system. The main disadvantage of the numerical sizing method is it needs a long computation time to identify one optimal configuration based on a predetermined reliability level [29]. It needs a long time because it computes the reliability of the system for all configurations of design space. In the analytical method, equations for the PV system size in terms of system reliability can be developed and utilized. With this method, the calculation of a system's size is very simple and accurate, but the complexity of deriving these equations' coefficient is the main drawback of this method.

Anyway, many issues should be taken into consideration when sizing a PVPS [30]. These issues are accurate information about the water flow rate profile demand, the efficiency and specification of the PVPS components,

Table 2.1 Summary of reliability, feasibility, and performance of PVPS.

Authors	System efficiency	Average flow rate	System size	Climate nature	Water head	Productivity rate per year
Alajlan et al. [1]	–	600 L/h	0.98 kWp	Saudi Arabia	50 m	2681.655 m^3/kWp
Posorski [3]	3%–5%	–	2 kWp	Seven different countries	5–125 m	–
Metwally et al. [4]	–	150 m^3/day	6.048 kWp	Egypt	30 m	9052.730 m^3/kWp
Arab et al. [5]	1.5%–2%	6 m^3/day	(4S × 3P), (3S × 4P) BP Saturn	Tamanrasset, Oran, Algiers and Bechar, Algeria		–
Belgacem [7]	2.5%	6.5–30 m^3/day	2.1 kWp	Kairouan, Tunisia	65–112 m	3172.215 m^3/kWp
Boutelhig et al. [8]	≤3.8%	4–6 m^3/day	–	Ghardaia, Algeria	19–35 m	–
Benghanem et al. [9]	–	≤ 22 m^3/day	(6S × 3P), (12S × 2P), (8S × 3P), (6S × 4P) × 75 Wp	Madinah, Saudia Arabia	80 m	4461.030 m^3/kWp
Mokeddem et al. [10]	–	6–65 L/min	1.5 kWp	Oran, Algeria	0.6 m and 11 m (static head)	6219.600 m^3/kWp
Al-Smairan [11]	–	45 m^3/day	5.9 kWp	Tall Hassan, Jordan	105 m	2783.855 m^3/kWp
Vick et al. [12]	5% (diaphragm) 7% (helical)	0–14 L/min 0–8.4 L/min	Different sizes	Amarillo, Texas, United States	Different heads	–
Kaldellis et al. [13]	5%	20 m^3/day	(2S × 6P) × 51 Wp	Greece	30 m	11,928.200 m^3/kWp
Senol [14]	–	18 m^3/day and 52 m^3/day	2 modules × 230Wp and 8 modules × 200Wp	Turkey	20 m	14,282.450 m^3/kWp 11,862.500 m^3/kWp

Continued

Table 2.1 Summary of reliability, feasibility, and performance of PVPS—cont'd

Authors	System efficiency	Average flow rate	System size	Climate nature	Water head	Productivity rate per year
Kolhe et al. [15]	3% (theoretical) 2.75% (exp.)	$38\,m^3$/day	(4S × 5P)	New Delhi, India	5 m	–
Setiawan et al. [16]	–	1.44–3.24 m^3/h	3.2 kWp (8S × 4P)	Purwodad, Indonesia	218.34 m	3202.875 m^3/kWp
Hamrouni et al. [17]	3%	7.7–14.7 m^3/day	2.1 kWp (14S × 3P)	Louata, Tunisia	65 m	1946.545 m^3/kWp
Benghanem et al. [18]	3.26%, 3.44%, 3.79%, and 4.41%	$\leq 33\,m^3$/day	1.8 kWp (8S × 3P)	Madinah, Saudia Arabia	50, 60, 70, and 80 m	6691.545 m^3/kWp
Mohammedi et al. [20]	1.299%[a] 0.83%[b]	43.66–27.9 L/min	990Wp (6S × 3P)	Bejaia, Algeria	11 m	9498.030 m^3/kWp
Cloutier et al. [21]	–	10, 20, and 30 m^3/day	840, 1680 and 2520 Wp	Nigeria	40 m	4345.325 m^3/kWp

[a]Without shading.
[b]With shading.

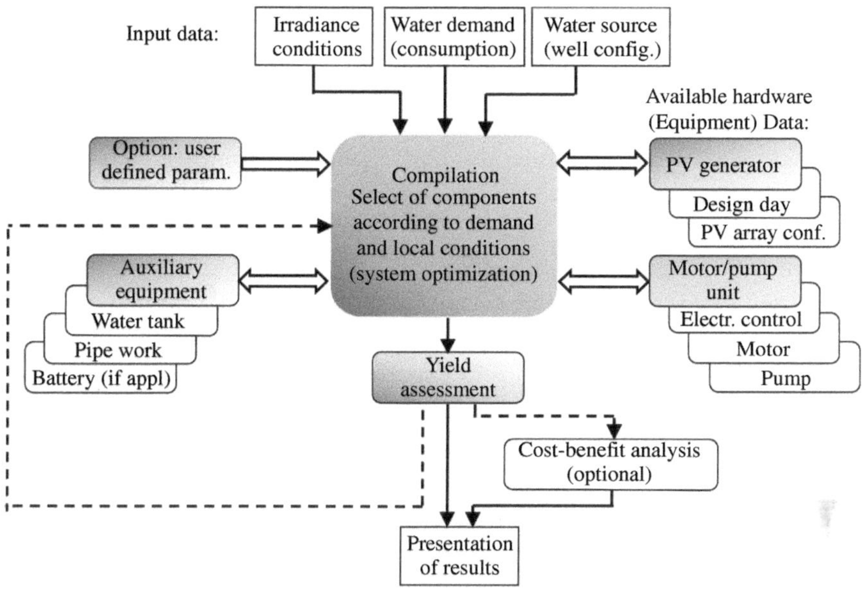

Fig. 2.2 Design parameters of a PV-water pumping system.

the total dynamic head, the pumping time factor, the solar radiation, the PV array power, and the capacity of the storage tank. Fig. 2.2 illustrates the parameters that affect the design of a PVPS.

In this section, research related to PVPS sizing is reported. Moreover, the reviewed methods are summarized at the end of the section so as to provide a clearer comparison. From the literature, the three aforementioned sizing methods are used to design a PVPS. The intuitive methods have been used and applied for many PVPSs. In [31], a tool for designing a water pumping system is developed. The design tool aims to match the produced and demanded water. The system is simulated based on specific climatic data. The size of the PV array is then calculated based on the total system efficiency, the solar energy, and the daily hydraulic energy. Similarly, in [32], two simplified design procedures, namely peak sun hours and hydraulic energy rule of thumb, are reviewed and compared. The month with the least solar radiation and the greatest water demand is considered as the design month to specify the required size of the PV array. Furthermore, in [33], the authors presented a design methodology for a photovoltaic water pumping system. The design procedure starts with the calculation of the solar radiation on tilted PV arrays with an optimal tilted angle to collect the largest amount of solar energy. After that, the optimal sizing of the PV arrays, the inverter, the batteries, the cable, and the motor pump set are done based on the worst month and peak sun hours (PSH). An economic analysis based on the life cycle cost (LCC) method is carried out for three systems: the PV, the grid, and the DG-based systems. According to the results, the system powered by a PV array was the most cost-effective as compared to the other two conventional power sources. However, 38% of the produced energy by the PV array was not used. In Ref. [34], a simple sizing method based on standard graphical performance information provided by the manufacturer is proposed to size a PVPS. The authors use a hypothetical constant called an equivalent total head instead of the combination of vertical and head loss. They used DASTPVPS software to experimentally derive an equation that is correlated to the equivalent total head with borehole characteristics. The results showed that the inclusion of an equivalent total head equation boosts the performance of the system and the design accuracy.

On the other hand, the numerical method is used as well for many PVPS. In [35], a methodology for sizing a PV array to meet both water and electricity demands is proposed. The methodology consists of two analytical steps. In the first step, experimental data are utilized to compute the area and configuration of the PV array and the maximum capacity of the storage battery of a system that meet the desired energy demand. On the other hand, in the second step, the authors simulated the proposed configurations to investigate the energy behavior of the system over a specific time period. A PV size versus maximum battery capacity curve is obtained from the numerical algorithm and based on that, the optimal design is chosen. The optimum configuration is selected based on the minimum system cost achieved. In Ref. [36], a numerical method for sizing a PVPS is presented. The optimization aims to minimize system cost subject to a specific reliability. The method depends on the loss of power supply probability (LPSP) concept. It uses hourly solar radiation and ambient temperature data for 1 year's time. A constant load profile is used in this work with differenet head levels. Four values of LPSP are used in the simulation: 0, 0.01, 0.05, and o.1. The simulation program provides several configurations that meet the desired reliability level (LPSP). After that, the optimal configuration that investigates the lowest cost is selected. In [37], an optimal sizing method for a PVPS based on an LPSP is proposed. The LPSP is used to specify the reliability of a set of system configurations that meet the desired load demand. After that, an economic evaluation is applied to these configurations to find the optimal configuration that achieves the minimum cost at the desired reliability.

In general, the drawback of the numerical method is the need for a long time to simulate the performance of the system over a wide range of configurations. Therefore, some authors use heuristic techniques to size PVPSs [38–42]. Tao Ma et al. [38] proposed modeling and designing procedures for PVPS. The authors presented an optimization of the system configuration based on zero load rejection (zero LPSP). The optimum system is defined as the system that investigates the desired reliability level at minimum cost. The genetic algorithm (GA) technique is used to solve this optimization problem. In [39], an optimization of a small photovoltaic water pumping hydro energy storage system for a small village in north Nigeria utilizing a particle swarm optimization (PSO) algorithm is proposed. The PSO algorithm is used to minimize the cost of the system and optimize the sizes of the devices based on different constraints such as weather condition, user requirements, and operation strategy. The authors in [40] used a hybrid sizing method that combines the numerical and heuristic techniques to size a hybrid PV-DG powered pumping system that is used to irrigate Mediterranean crops. In [40], a possible design space that contains system configurations meeting the desired system reliability is generated on the basis of a numerical method. A GA was then used to select the configuration that investigates the minimum lifetime cost. Muhsen et al. [41] proposed a differential evolution-based multiobjective optimization algorithm to optimally size a PVPS. In [41], three objective functions–loss of load probability, life cycle cost, and excess water volume–are aggregated by a single objective function based on predetermined weights. The main drawback of [41] is the complexity of initializing the weights used by the aggregation function. Furthermore, the proposed method in [41] converts the multiobjective optimization to a single objective optimization; therefore, the sizing method identifies only a single configuration at each run time. Muhsen et al. [42] proposed a differential evolution based multiobjective optimization (DEMO) algorithm to size a standalone PVPS as well as to overcome the drawbacks of [41]. In [42], the PVPS is sized based on a technical criterion (loss of load probability) and an economic criterion (LCC). The main drawback of DEMO [42] is that the computation complexity is increased by increasing the number of criteria that are considered to size the PVPS.

As for the analytical method, it is employed to size a PVPS in many studies as well. In [43], a method for calculating the PVPS loss of load probability (LLP) is presented. The proposed method aims to describe the ratio of PV maximum current to the nominal motor current ($I_{mp}/I_{n.mot}$) as a function of the ratio of PV maximum voltage to the motor nominal voltage ($V_{mp}/V_{n.mot}$) at different values of the LLP. The resulted ($I_{mp}/I_{n.\ mot}$)-($V_{mp}/V_{n.\ Mot}$) curve divides the plotting plane into two distinct regions called the oversizing region (above the curve) and the undersizing region (below the curve). The motor pump is characterized by *I*-*V* and *Q*-*V* curves. These curves are obtained by experiment, and then these curves are fitted by linear and polynomial equations. In Ref. [44], an analytical method to optimize the size of the PVPS is proposed. The method is based on metorological data, water demand, well depth, simple modeling, and curve fitting techniques. The authors used the speed of pump rotation and the affinity law to find the flow rate of PVPS in terms of speed and head by a two-variable polynomial curve fitting function. After that, the curve fitting method is also used to obtain a correlation for the pump efficiency in terms of the flow rate. In Ref. [45], a method for PVPS sizing based on the loss of load probability (LLP) is proposed. The study investigated the effect of total head, peak power of the PV array, water consumption, and storage tank capacity on the performance of the system. The authors proposed a direct relationship between the input power and the flow rate based on experimental data. According to the simulation results, the peak power of the PV array and the total pumping head are strongly related to the relative deficit of the system. On the other hand, the capacity of the storage tank and the profile of water consumption have a minor effect on the relative water deficit with a given total pumping head. In [46], a comparative study of PVPSs is done to select an optimum configuration of the system. The optimum configuration of the PV array is selected based on an experimental performance of four systems that have 2P × 2S, 2P × 1S, 1P × 2S, and 1P × 1S configurations that power a 120 W/24 V DC Shurflo pump. The experimental performance is monitored for three different heads (10 m, 15 m, and 25 m). According to the results, the authors recommended combining the 2P × 1S and 1P × 2S PV array configurations together to power the pump. The combination is controlled by an electronic PV array reconfiguration controller to activate the 2P × 1S PV to power the system in the morning so as to supply the system with an adequate starting current. The 1P × 2S PV array is used to power the system during optimum sunlight. In [47], an approach to optimize and improve the performance of a PVPS is proposed. The approach is represented by optimizing the PV array configuration based on solar radiation. Three PV configurations are used in this research: 4P × 1S, 2P × 2S, and 1P × 4S. These configurations are chosen to address the low, medium, and high solar radiation intensity conditions, respectively. The proposed approach provides a sufficient starting current to the motor, especially in the morning, evening, and cloudy weather. Finally, Table 2.2 shows a summary of different sizing techniques proposed for PVPS.

Table 2.2 Summary of sizing of PVPS.

Authors	Sizing methodologies	Outcome
Campana et al. [30]	Intuitive method (the worst month method)	Used the lowest ratio of monthly and daily average of radiation to water demand and initial investment cost to specify the optimal and cost effective PVPS design. The results show that the fixed PV array-driven AC pump is the most cost effective type
Caton [31]	Intuitive method (the worst month method)	The least solar radiation and greatest water demand month are considered to specify the optimal size of the PV array
Ebaid et al. [32]	Intuitive method (the worst month method)	The system is oversizing, where 38% of the PV producing energy is excess because the authors used a simple and inaccurate sizing method
Narvarte et al. [33]	Analytical method	Sizing the PV based on the standard graphic provided by the PV pump manufacturer and experimental equivalent total head equation
Hadj Arab et al. [42]	Analytical method	For a given PV array size, the LLP sensitivity is varied according to the location of the system. Also, the proposed method can be represented as a graphical tool for sizing the PVPS for a given head
Kalddlis et al. [34]	Numerical method	Obtained the optimum PV size and maximum battery capacity, according to a numerical algorithm and economic analysis
Bouzidi [35]	Numerical method	For a given reliability, realize an optimum size of the pumping system with a minimum cost using a numerical method based on the LPSP
Bakelli et al. [36]	Numerical method	Realize an optimum configuration (optimum number of PV modules and capacity of storage tank based on technical (LPSP) and economic (LCC) optimization criteria)
Ma et al. [37]	AI method-based genetic algorithm	For zero reliability and a minimum cost, an optimum system configuration (number of PV modules and capacity of the upper reservoir) is specified

Continued

Table 2.2 Summary of sizing of PVPS—cont'd

Authors	Sizing methodologies	Outcome
Stoppato [38]	AI method-based PSO algorithm	The PSO algorithm is used to optimize the system technically and economically. The systemis capable of covering the water demand and 9% of the electricity demand
Carroquino [39]	AI method-based genetic algorithm	Optimize the size of the system numerically subject to a desired *LLP*; afterward the GA is used to optimize the system based on *LCC*
Dhiaa [40]	AI method-based DE algorithm	Optimally sized PVPS using differential evolution-based multiobjective optimization algorithm. The multiobjective optimization problem is converted to a single objective problem by aggregating the objectives LLP, LCC, and excess water in a single objective function based on predetermined weights
Dhiaa [41]	AI method-based DEMO algorithm	DEMO is proposed to size PVPS by minimizing two objectives (LLP and LCC) simultaneously. A set of PVPS configurations is obtained that constitute a trade-off between the objectives (Pareto front)

2.5 Photovoltaic water pumping system control procedures

In general, it is important to control a PVPS optimally so as to achieve optimal operation of the system and consequently a reliable system. Many control approaches have been developed by researchers to efficiently operate a PVPS [48–54]. These approaches include MPPT algorithms, voltage regulation, frequency control, and load matching.

MPPT is a technique applied to PV systems to extract and track the maximum power generated by the PV array at different weather conditions. The MPPT algorithms can be classified into conventional algorithms (normally effective in case of not having any shading objectives) and algorithms that are based on stochastic and AI techniques. In general, the conventional algorithms may fail to track the MPP under partial shading conditions and rapidly changing solar radiation [55]. Conventional MPPT algorithms include perturb and observation (P&O), incremental and conductance (InCond), hill climbing (HC), fractional V_{oc}, fractional I_{sc}, ripple correlation control, and load current or load voltage maximization. The P&O and InCond algorithms are the most widely used in PV systems [55]. The concept of a P&O algorithm is based on perturbing the output voltage of a PV array with a small value, and then according to the direction of change in PV power, the sign of the next perturbation value is specified. Meanwhile, in the InCond algorithm, the PV voltage is adjusted based on the comparison between the instantaneous conductance (I/V) and the incremental conductance($\Delta I/\Delta V$).

In [56], a nonlinear control technique that depends on feedback linearization control is used to enable the system to operate at MPP by adjusting the duty cycle ratio of the buck-boost converter. The concept of feedback input-output linearization control in [56] is based on differentiating the PV output voltage to find a direct relationship with a duty cycle ratio. The duty cycle ratio (u) is the control variable that is the output of a nonlinear controller while the solar radiation (G), temperature (T), motor armature current (I_a), PV output voltage (V_P), and output of the linear P-controller (v) are the inputs of the nonlinear controller. Meanwhile, the input of the P-controller is the difference between the PV output voltage and the reference voltage (V_{ref}), which corresponds to the voltage at the maximum power point. Similarly, in [57], the InCond method is utilized to track the MPP of the PV array in a PVPS with minimum losses point tracking (MLPT) for the induction motor. The proposed control strategy improved the system's efficiency and reduced the losses of the motor. The experimental results indicated that 8% of the input power is saved by reducing the motor losses.

In Ref. [58], an MPPT algorithm and sun-tracking algorithm were applied to improve the overall efficiency of a PVPS that operates with a different total head ranging from 5 to 35 m. The MPPT technique proposed in [58] is based on the difference between the measured voltage of the PV array and a reference voltage at which the maximum power is obtained. The difference is compared with a sawtooth signal via a comparator to produce the switching signal to control a MOSFET switch. The duty cycle of the triggering signal depends on the output voltage of the PV array and the type of load. On the other hand, the proposed sun tracker for PVPS increased the PV output power by 36%. In [59], two approaches of P&O-based MPPTs are proposed for PVPS. These approaches are the direct duty cycle-based P&O and the reference voltage-based P&O. The effect of the algorithm's parameters (step size and perturbation rate) on the performance of the system is investigated. Furthermore, the drawbacks and advantages of each approach under different weather conditions are demonstrated. According to the results, the reference voltage perturbation has a faster response than the direct duty ratio perturbation at rapidly changing solar radiation and temperature. Meanwhile, the stability characteristics of the direct duty ratio perturbation are better than the reference voltage perturbation technique, particularly with high perturbation. In addition, the direct duty cycle ratio perturbation technique offers higher energy utilization efficiency than the reference voltage perturbation technique for both slow and rapidly changing solar radiation. In [60], the equation of the terminal voltage of the PV array and the equation of the derivative of power with respect to current at the maximum power point are solved together by a numerical method for maximum voltage and current. After that, a curve-fitting approach is used to describe the current and voltage at the MPP in terms of the photocurrent and PV open circuit voltage, respectively, for all levels of solar radiation. In [61], an electronic circuit is used to produce a fixed duty cycle ratio for the step-up converter to enable the PV array to operate at MPP regardless of solar radiation variations. The quality of matching DC-motor drive with volumetric/centrifugal pump that is directly coupled with a photovoltaic is demonstrated by comparing the *I*-*V* characteristics of the PV array and the motor pump. SUPER SCEPTER computer software is used to obtain the *I*-*V* characteristics of the motor-pump set. The results show a large mismatch between the photovoltaic MPP and the equilibrium operating point of the DC motor coupled with a photovoltaic pump. On the other hand, the DC motor-driven centrifugal pump is more compatible with a photovoltaic array. In [62], matching between a photovoltaic and a three-phase induction motor for optimal operation using an enhanced PV model is presented. A PCU that consists of a double step-up converter and a six-step voltage source inverter is used to track the MPP of the PV array in the system.

In Ref. [63], an online fuzzy controller is proposed to adjust the duty cycle ratio of a DC-DC converter to optimize the global efficiency of a PVPS by maximizing the input power of a centrifugal pump, and consequently the water pumped flow rate is maximized. The inputs of the proposed fuzzy controller are the variation of both the global efficiency and the chopper ratio. Meanwhile, the output of the fuzzy controller is the adjusted duty ratio. The output and input scaling factors of the controller are chosen based on the trial-and-error method. Five triangle membership functions are chosen to express the output and input variables of the controller. In Ref. [64], a feed-forward artificial neural network (ANN)-based controller for tracking the optimal operating point for a PV water pumping system is proposed. The optimal operating point is either the maximum PV power point (MP) or the gross mechanical energy (GME) operating point of the pumping system. The proposed ANN controller adjusts the duty ratio of the buck-boost converter based on the solar radiation to match between the PV array and the motor-pump set. Thus, the input and output of the ANN are the solar radiation and chopping ratio corresponding to the GME or MP point, respectively. The ANN comprises one hidden layer with a bipolar sigmoid activation function as well as an output layer with a linear activation function. In Ref. [65], a technique based on fuzzy logic control is proposed to identify the MPP by adjusting the duty cycle ratio of the DC-DC converter. The method of tracking the MPP is based on the ratio of PV power variation to the PV current (dP/dI), where this value represents the error that should be minimized to be close to zero. The input variables of the fuzzy controller are the error and the derivative of error, where each variable is represented by five triangle membership functions. Meanwhile, the duty cycle ratio is the output of the controller, which is expressed by seven singletons membership functions. In Ref. [66], an MPPT for PVPS is presented. The proposed adaptive controller is based on nonlinear autoregressive moving average-L2 (NARMA-L2) ANNs, which comprise two feed-forward neural networks. The inputs of NARMA-L2 are the motor input power and the maximum PV power as the reference power. Meanwhile, the duty cycle ratio of the maximum power point data are used to train the network. The simulation results show that the proposed controller has a good performance and fast response. In Ref. [48], a single-stage Z-source inverter with an MPPT based on the fuzzy logic-incremental conductance technique is applied to a PVPS to extract and track the maximum power point of the PV array. A direct torque control strategy is used to drive the brushless DC motor coupled with a centrifugal pump. A particle swarm optimization is proposed to regulate the parameter of the PID-speed controller of the motor to improve the transient response. Simulation results indicate the effectiveness of the proposed control strategies.

In Ref. [67], the authors developed two control strategies to improve the efficiency of PVPS. The first strategy is based on a controller that consists of an InCond-MPPT algorithm that estimates PV voltage at the MPP. The role of the PI-controller is to produce the ratio vector of the PWM of the inverter. On the other hand, the second strategy is based on the direct torque control (DTC) to control the flux and torque of the induction motor by selecting the optimum switching vector for the inverter. The main drawback of the first strategy is that the pump has not pumped water in spite of the PV generator operating at MPP during the low solar radation hours. Meanwhile, the main drawback of the second strategy was the significant ripple in the flux and torque. According to the results, the second control strategy offers a fast response without overshoot and fewer steady-state oscillations under rapid solar radiation and temperature changes. Furthermore, the adoption of the DTC strategy assists in overcoming and eliminating the complexity of using the MPPT algorithms.

In Ref. [68], the authors proposed control strategies for two types of PVPSs based on the utilized motor in the system. The first system utilized a DC motor while the second system utilized an induction motor. In the first case, an approach to match between the PV array and motor to achieve maximum mechanical power is presented. The maximum mechanical power is extracted by controlling the excitation of the motor. On the other hand, in the second case, the maximum power is obtained by controlling the frequency of the voltage source inverter that connects both the PV array and the induction motor. The results show that the maximum mechanical power extracted from the separately excited DC motor is higher than that in an induction motor. However, the DC motor utilizes the power of the PV array better than the induction motor. In Ref. [69], two strategies for controlling PVPSs are proposed. The first control strategy is based on the concept that the PV array operates at maximum power points with a variable water discharge rate. The second control strategy is based on the maximum efficiency of the motor with a unique motor speed that is related to the optimum motor efficiency and limited water discharge rate. Table 2.3 illustrates a brief comparison between the two control strategies.

In Ref. [70], an approach is developed to match between the PV array and the separately excited DC motor-pump set for the maximum power operation. The approach is based on finding an expression for the optimum motor field current in terms of the motor-pump parameters and PV array MPP. The operating point of the motor-pump follows the maximum power trajectory of the PV array by adjusting the field current of the DC motor according to the solar radiation. The results show that the required motor excitation current is equal to the rated excited current in peak irradiance. At low irradiance, the DC motor needs double the rated field current to match the MPP of the PV array. In [71], two control strategies are proposed to match between the input characteristics of the DC motor and the PV

Table 2.3 Brief comparison between two control strategies.

First strategy	Second strategy
1. The voltage-to-frequency ratio (v_s/ w_s) is constant while the system's frequency is adjusted **2.** The operation of the pumping system begins as the level of irradiance is 25% of the maximum value; therefore, the volume of water pumped is greater than that in the second strategy **3.** The system operates at the maximum power points of the PV array **4.** The water discharge rate is variable **5.** A closed loop control is used to implement the strategy, where the voltage and speed of the motor are sensed to adjust the output frequency via changing the triggering angle of both the DC-DC converter and inverter to keep the voltage-to-frequency ratio constant	**1.** The voltage-to-frequency ratio (v_s/w_s) is adjusted while the frequency is constant **2.** Operation of the pumping system begins as the level of irradiance is 50% of the maximum value. So, the amount of the pumped water is less than that in the first strategy **3.** The motor operates at maximum efficiency and the PV array operating characteristics are controlled to match the operation point of the motor at different irradiance **4.** The water discharge rate is limited because in this strategy, the induction motor operates at a unique speed corresponding to the maximum motor efficiency **5.** The closed loop system senses the motor voltage to dominate the motor torque via adjusting the output voltage of the DC-DC converter. The speed regulation of the motor can be improved by sensing the actual speed

output characteristics by adjusting the duty cycle ratio of the buck-boost converter. The first strategy aims to extract the maximum power of the PV array by deriving a direct correlation of the duty cycle ratio. Meanwhile, the second strategy aims to maximize the gross mechanical energy of the system per day by deriving a correlation between the duty cycle ratio and the optimal power value, which is related to the maximum power of the PV array. It can be observed from the steady-state simulation results that the chopping ratio and armature voltage for both types of motor in the second strategy operation case are higher than in the first strategy operation case. Meanwhile, the armature current in the first strategy operation is higher than the second strategy operation along the solar radiation variations. Therefore, the copper losses of a DC motor in the second strategy operation are less than the losses during the first strategy operation. Consequently, the second strategy utilization improves the efficiency of a DC motor and the whole pumping system. In [72], two control approaches are proposed for a PVPS. The first approach is for speed control while the second is for hysteresis current control. Traditional and fuzzy logic (FL)-based PI controllers are used in these approaches. The simulation results for the system without the proposed controllers show high fluctuations in the torque and phase current of the motor. Meanwhile, the simulation results of the system with the traditional PI controllers indicated a little ripple in both the torque and current of the motor, but a big overshoot is clear in the transient response of the motor. On the other hand, the dynamic response of the system with an FL controller is improved whereas the ripple in the torque response is eliminated. Furthermore, the FL controller is capable of quickly converging the speed of the motor to the reference value with zero steady-state error and without overshoot. In Ref. [73], the transient performance of the system is dominated by two sets of differential equations based on the status of the chopper switch. The two sets of differential equations are solved in MATLAB using a fourth-order Runge Kutta numerical method. The authors used a boost converter to match between the motor-pump set and the PV array to increase the flow rate of the pumped water. Table 2.4 shows a summary of all the aforementioned control methods applied to PVPS.

Table 2.4 Summary of control strategies of PVPS.

Authors	Approach of control	Notes
Andoulssi et al. [55]	MPPT	Provide an MPPT based on a nonlinear control technique with feedback linearization control to adjust the duty ratio of the buck-boost converter
Corrêa et al. [56]	MPPT and MLPT	The InCond algorithm and MLPT for the induction motor are used to improve the efficiency of the system and to reduce the motor losses
Katan et al. [57]	MPPT	An MPPT technique and sun tracking control system are used to improve the overall system efficiency
Elgendy et al. [58]	P&O MPPT	A comparison between the direct duty cycle perturbation and the reference voltage perturbation algorithms is investigated

Table 2.4 Summary of control strategies of PVPS—cont'd

Authors	Approach of control	Notes
Alghuwainem [59]	Load matching	According to the author, there is a unique and the optimum duty cycle ratio for a step-up converter at which the maximum power transfer from the PV array to the motor occurs at all levels of radiation
Applebaum [60]	Load matching	The author has calculated that a DC motor-driven centrifugal pump is more compatible with a photovoltaic array
Akbaba [61]	Load matching	The PCU stimulates the PV array to operate on the maximum power trajectory
Benlarbi et al. [62]	Online optimization of the global efficiency	An online fuzzy controller is proposed to adapt the electrical load impedance of a PV array by adjusting the chopper ratio based on the maximum global efficiency of the system at different radiation levels
Veerachary et al. [63]	MPPT	An ANN controller is used to adjust the duty ratio of the buck-boost converter according to the solar radiation for tracking the optimal operating point
Mazouz et al. [64]	MPPT	A fuzzy controller is used to choose the appropriate duty cycle of the DC-DC converter to enable the system to operate at MPP
Kassem [65]	MPPT	A NARMA-L2 ANN adaptive controller is proposed to adjust the duty ratio of the buck-boost converter
Mozaffari Niapour et al. [66]	MPPT & DTC	This proposes an MPPT algorithm based on a fuzzy logic-incremental conductance technique, also a DTC strategy is used to drive the motor of the PVPS
Caton P. [31]	MPPT	Uses a commercial MPPT to match between the PV modules and the motor-pump set

Continued

Table 2.4 Summary of control strategies of PVPS—cont'd

Authors	Approach of control	Notes
Hamrouni et al. [17]	MPPT and *V/f* law	The InCond-MPPT algorithm with the *V/f* law is used to control and extract maximum PV power
Zaki et al. [67]	Frequency and excitation control	Maximum mechanical power to drive a centrifugal pump by a DC motor and an AC motor is attained by controlling the excitation and the frequency of the voltage source inverter for both the DC and AC motors, respectively
Eskander et al. [68]	Load matching	Two control strategies are used; the first is the PV operates at maximum efficiency with a variable flow rate while the other strategy is the induction motor operates at maximum efficiency with a limited flow rate
Akbaba et al. [69]	Load matching	The matching between the PV array and the separately excited DC motor is achieved by adjusting the motor field current and represented in terms of the motor pump parameters and the MPP characteristics of the PV array
Chary Mummadi [70]	Load matching	Stimulate the PVPS to operate at MPP or at the maximum daily gross mechanical output energy (GME) by adjusting the duty ratio of the buck-boost converter
Terki et al. [71]	PI-controller	A fuzzy PI speed controller and hysteresis current controller are proposed to control the speed and current loops, respectively
Akbaba et al. [72]	Load matching	The step-up converter increases the motor output power by around 15%
Dubey et al. [73]	Voltage regulator	Two controllers are used; one to control the output voltage of the DC-DC converter while the other controller based on a vector oriented control scheme controls the voltage source inverter

2.6 Load matching evaluation technique

The load-matching factor is usually used to evaluate the performance of PV water pumps. This approach has been the most effective and simplest in analyzing the performance of heat collectors during the past few decades. The same approach can be adapted for PV system performance evaluation. The load-matching factor is the ratio of energy acquired by the hydraulic load to the maximum PV array power extracted in a 1-day period. This could be stated as the ratio of the power that the PV array delivers to the motor-pump subsystem during the day to the maximum electrical power that can be obtained from the array throughout the day. This can also be defined as the ratio of the actual array output used for water pumping to the array output capability as below,

$$\Phi = \frac{\text{Actual array output used for water pumping}}{\text{Array output capbility}} \tag{2.1}$$

This parameter is very useful for assessing how much of a PV array's real power production capability is being used.

Load-matching factor can be determined from theoretical models or from field experimental data (from solar radiation and PV array power output data). Theoretically, the load-matching factor is expressed as:

$$\Phi = \frac{P_h}{P} \tag{2.2}$$

where P is the PV array power output, and Ph is the power acquired by the hydraulic load.

Load-matching factor can also be determined graphically from experimental data. Fig. 2.1 represents the typical daily average and hourly PV array power output curve. The horizontal line (CD) shows the threshold level of the electrical load on the PV pump. If we take a short time interval (the shaded areas A and B), the total array output capability can be obtained by integrating over a given time. The actual array output used for pumping is then the area of the curve above the threshold line. The area below the threshold line is the power wasted due to friction and other losses. Therefore, the load matching factor is the ratio of the curve area above the threshold line to the total area of the curve.

Similarly, the load-matching factor can be determined from the daily average and hourly solar radiation curve. Because the PV array power output depends on the distribution of solar radiation, the solar radiation curve over

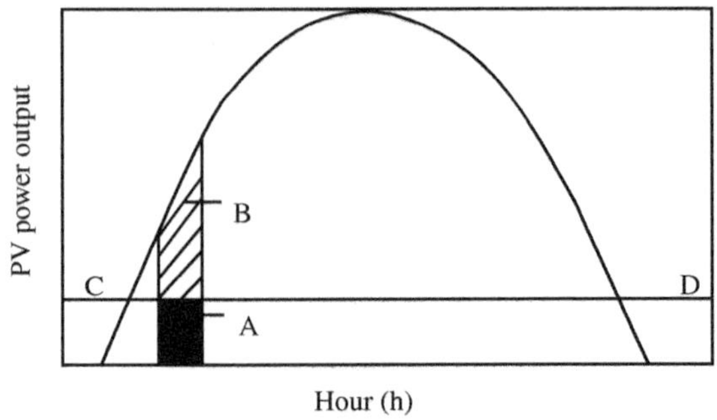

Fig. 2.3 A typical daily average hourly PV array power output.

the day will have a similar shape to Fig. 2.3, and the load-matching factor from the solar radiation curve is defined by

$$\Phi = \frac{\sum^{N}(G_r - G_c)}{\sum^{N} G_r} \tag{2.3}$$

where N is the number of daily hour readings, G_c is the critical solar radiation (or the threshold) to start water pumping, and G_T is the solar radiation on the tilted surface of the PV array.

The load-matching factor is usually easier to estimate from the PV array power output than from the solar radiation curve. This is because the critical PV array power output is constant, unlike the critical line on the solar radiation curve, which may not be a straight line. The optimum load-matching factor of this PV pump, estimated from the power output curve, is 0.84. This factor could be a bit higher if the trees did not shade the PV array. This figure is still quite high and shows that the system components are well matched, or properly configured. The threshold line shifts up or down depending on the type of motor-pump subsystem, capacity, and weather conditions. Optimum load matching of the PV water pumping systems mainly depends on solar radiation and load profiles. Because the load-matching factor is the ratio of the curve's area above the threshold line to the total area of the curve, the load-matching factor for a given PV pump is higher on good solar radiation days than during low solar radiation days. Therefore, the load-matching factor varies from season to season. To maximize the performance of a PV water pumping system, the operation of the load line must be close to the maximum PV array power line throughout the day. Thus, because the loadmatching factor is a measure of design quality of the system within the local environment, it has to be as high as possible for all seasons. It is also possible to achieve a higher loadmatching factor by carefully selecting the proper size of the PV array and the motor-pump subsystem at the best solar radiation locations. The load-matching factor can be close to unity for a well-matched PV pumping system on the best solar radiation day.

The simplest method of measuring a PV pump's performance is to take daily readings of the solar radiation on the PV array's plane, the daily volume of water pumped, and the static head of the well. This allows the hydraulic energy to be calculated, and then the system efficiency can be estimated at different solar radiation energy levels.

References

[1] Alajlan SA, Smiai MS. Performance and development of PV - plant for water pumping and desalination for remote area in Saudi Arabia. Renew Energy 1996;8:441–6.
[2] Bucher W. Aspects of solar water pumping in remote regions. Energy Sustain Dev 1996;3:8–27.
[3] Posorski R. Photovoltaic water pumps, an attractive tool for rural drinking water supply. Sol Energy 1996;58:155–63.
[4] Metwally HMB, Anis WR. Performance analysis of PV pumping systems using switched reluctance motor drives. Energy Convers Manag 1997;38:1–11.
[5] Arab AH, Chenlo F, Mukadam K, Balenzategui JL. Performance of PV water pumping systems. Renew Energy 1999;18:191–204.
[6] Hammad MA. Characteristics of solar water pumping in Jordan. Energy 1999;24:85–92.
[7] Belgacem BG. Performance of submersible PV water pumping systems in Tunisia. Energy Sustain Dev 2012;16:415–20.
[8] Boutelhig A, Hadjarab A, Bakelli Y. Comparison study to select an optimum photovoltaic pumping system (PVPS) configuration upon experimental performances data of two different DC pumps tested at Ghardaïa site. Energy Procedia 2011;6:769–76.
[9] Benghanem M, Daffallah KO, Joraid AA, Alamri SN, Jaber A. Performances of solar water pumping system using helical pump for a deep well: a case study for Madinah, Saudi Arabia. Energy Convers Manag 2013;65:50–6.
[10] Mokeddem A, Midoun A, Kadri D, Hiadsi S, Raja IA. Performance of a directly-coupled PV water pumping system. Energy Convers Manag 2011;52:3089–95.
[11] Al-Smairan M. Application of photovoltaic array for pumping water as an alternative to diesel engines in Jordan Badia, Tall Hassan station: case study. Renew Sust Energ Rev 2012;16:4500–7.
[12] Vick BD, Clark RN. Experimental investigation of solar powered diaphragm and helical pumps. Sol Energy 2011;85:945–54.
[13] Kaldellis JK, Meidanis E, Zafirakis D. Experimental energy analysis of a stand-alone photovoltaic-based water pumping installation. Appl Energy 2011;88:4556–62.
[14] Senol R. An analysis of solar energy and irrigation systems in Turkey. Energy Policy 2012;47:478–86.
[15] Kolhe M, Joshi JC, Kothari DP. Performance analysis of a directly coupled photovoltaic water-pumping system. IEEE Trans Energy Convers 2004;19:613–8.
[16] Setiawan AA, Purwanto DH, Pamuji DS, Huda N. Development of a solar water pumping system in karsts rural area Tepus, Gunungkidul through student community services. Energy Procedia 2014;47:7–14.

[17] Hamrouni N, Jraidi M, Chérif A. Theoretical and experimental analysis of the behaviour of a photovoltaic pumping system. Sol Energy 2009;83:1335–44.
[18] Benghanem M, Daffallah KO, Alamri SN, Joraid AA. Effect of pumping head on solar water pumping system. Energy Convers Manag 2014;77:334–9.
[19] Mahmoud E, el Nather H. Renewable energy and sustainable developments in Egypt: photovoltaic water pumping in remote areas. Appl Energy 2003;74:141–7.
[20] Mohammedi A, Mezzai N, Rekioua D, Rekioua T. Impact of shadow on the performances of a domestic photovoltaic pumping system incorporating an MPPT control: a case study in Bejaia, North Algeria. Energy Convers Manag 2014;84:20–9.
[21] Cloutier M, Rowley P. The feasibility of renewable energy sources for pumping clean water in sub-Saharan Africa: a case study for Central Nigeria. Renew Energy 2011;36:2220–6.
[22] Yesilata FZA. Effect of solar radiation correlations on system sizing: PV pumping case. Renew Energy 2008;33:155–61.
[23] Samimi J, Soleimani EA, Zabihi MS. Optimal sizing of photovoltaic systems in varied climates. Sol Energy 1997;60:97–107.
[24] Mellit A, Kalogirou SA, Hontoria L, Shaari S. Artificial intelligence techniques for sizing photovoltaic systems: a review. Renew Sust Energ Rev 2009;13:406–19.
[25] Mohamed AF, Elarini MM, Othman AM. A new technique based on artificial bee colony algorithm for optimal sizing of stand-alone photovoltaic system. J Adv Res 2014;5:397–408.
[26] Khatib T, Mohamed A, Sopian K. A review of photovoltaic systems size optimization techniques. Renew Sust Energ Rev 2013;22:454–65.
[27] Khatib T, Ibrahim IA, Mohamed A. A review on sizing methodologies of photovoltaic array and storage battery in a standalone photovoltaic system. Energy Convers Manag 2016;120:430–48.
[28] Kaplani E, Kaplanis S. A stochastic simulation model for reliable PV system sizing providing for solar radiation fluctuations. Appl Energy 2012;97:970–81.
[29] Acakpovi A, Xavier FF, Awuah-Baffour R. Analytical method of sizing photovoltaic water pumping system. In: IEEE 4th International Conference in Adaptive Science & Technology (ICAST); 2012. p. 65–9.
[30] Campana PE, Li H, Yan J. Dynamic modelling of a PV pumping system with special consideration on water demand. Appl Energy 2013;112:635–45.
[31] Caton P. Design of rural photovoltaic water pumping systems and the potential of manual array tracking for a west-African village. Sol Energy 2014;103:288–302.
[32] Ebaid MSY, Qandil H, Hammad M. A unified approach for designing a photovoltaic solar system for the underground water pumping well-34 at Disi aquifer. Energy Convers Manag 2013;75:780–95.
[33] Narvarte L, Lorenzo E, Caamaño E. PV pumping analytical design and characteristics of boreholes. Sol Energy 2000;68:49–56.
[34] Kaldellis JK, Spyropoulos GC, Kavadias KA, Koronaki IP. Experimental validation of autonomous PV-based water pumping system optimum sizing. Renew Energy 2009;34:1106–13.
[35] Bouzidi B. New sizing method of PV water pumping systems. Sustainable Energy Technol Assess 2013;4:1–10.
[36] Bakelli Y, Hadj Arab A, Azoui B. Optimal sizing of photovoltaic pumping system with water tank storage using LPSP concept. Sol Energy 2011;85:288–94.
[37] Ma T, Yang H, Lu L, Peng J. Pumped storage-based standalone photovoltaic power generation system: modeling and techno-economic optimization. Appl Energy 2015;137:649–59.
[38] Stoppato A, Cavazzini G, Ardizzon G, Rossetti A. A PSO (particle swarm optimization)-based model for the optimal management of a small PV(photovoltaic)-pump hydro energy storage in a rural dry area. Energy 2014;76:168–74.

[39] Carroquino J, Dufo-López R, Bernal-Agustín JL. Sizing of off-grid renewable energy systems for drip irrigation in Mediterranean crops. Renew Energy 2015;76:566–74.
[40] Muhsen DH, Ghazali AB, Khatib T. Multiobjective differential evolution algorithm-based sizing of a standalone photovoltaic water pumping system. Energy Convers Manag 2016;118:32–43.
[41] Muhsen DH, Ghazali AB, Khatib T, Abed IA, Natsheh EM. Sizing of a standalone photovoltaic water pumping system using a multi-objective evolutionary algorithm. Energy 2016;109:961–73.
[42] Hadj Arab A, Chenlo F, Benghanem M. Loss-of-load probability of photovoltaic water pumping systems. Sol Energy 2004;76:713–23.
[43] Boutelhig A, Bakelli Y, Hadj Mahammed I, Hadj AA. Performances study of different PV powered DC pump configurations for an optimum energy rating at different heads under the outdoor conditions of a desert area. Energy 2012;39:33–9.
[44] Salameh ZM, Dagher F. The effect of electrical array reconfiguration on the performance of a PV-powered volumetric water pump. IEEE Trans Energy Convers 1990;5:653–8.
[45] Martiré T, Glaize C, Joubert C, Rouvière B. A simplified but accurate prevision method for along the sun PV pumping systems. Sol Energy 2008;82:1009–20.
[46] Metwally HMB, Anis WR. Dynamic performance of directly coupled photovoltaic water pumping system using D.C. shunt motor. Energy Convers Manag 1996;37:1407–16.
[47] Arrouf M, Bouguechal N. Vector control of an induction motor fed by a photovoltaic generator. Appl Energy 2003;74:159–67.
[48] Moulay-Idriss C, Mohamed B. Application of the DTC control in the photovoltaic pumping system. Energy Convers Manag 2013;65:655–62.
[49] Raveendhra D, Joshi P, Verma RK. Performance and control system design for FPGA based CVMPPT boost converter for remote SPV water pumping system applications. In: Towards Sustainable Energy in Power and Energy Systems Conference; 2014. p. 1–6.
[50] Betka A, Moussi A. Performance optimization of a photovoltaic induction motor pumping system. Renew Energy 2004;29:2167–81.
[51] Firatoglu ZA, Yesilata B. New approaches on the optimization of directly coupled PV pumping systems. Sol Energy 2004;77:81–93.
[52] Hussein A, Hirasawa K, Hu J. A robust control method for a PV-supplied DC motor using universal learning networks. Sol Energy 2004;76:771–80.
[53] Kalogirou SA. Artificial neural networks in renewable energy systems applications: a review. Renew Sust Energ Rev 2001;5:373–401.
[54] Kamarzaman NA, Tan CW. A comprehensive review of maximum power point tracking algorithms for photovoltaic systems. Renew Sust Energ Rev 2014;37:585–98.
[55] Andoulssi R, Draou A, Jerbi H, Alghonamy A, Khiari B. Non linear control of a photovoltaic water pumping system. Energy Procedia 2013;42:328–36.
[56] Corrêa TP, Seleme Jr. SI, Silva SR. Efficiency optimization in stand-alone photovoltaic pumping system. Renew Energy 2012;41:220–6.
[57] Katan RE, Agelidis VG, Nayar C. Performance analysis of a solar water pumping system. In: Power Electronics; 1996.
[58] Elgendy MA, Zahawi B, Atkinson DJ. Assessment of perturb and observe MPPT algorithm implementation techniques for PV pumping applications. IEEE Trans Sustainable Energy 2012;3:21–33.
[59] Alghuwainem SM. Steady-state performance of DC motors supplied from photovoltaic generators with step-up converter. IEEE Trans Energy Convers 1992;7:267–72.
[60] Applebaum J. The quality of load matching in a direct-coupling photovoltaic system. IEEE Trans Energy Convers 1987;EC-2:534–41.

[61] Akbaba M. Matching induction motors to PVG for maximum power transfer. Desalination 2007;209:31–8.
[62] Benlarbi K, Mokrani L, Nait-Said MS. A fuzzy global efficiency optimization of a photovoltaic water pumping system. Sol Energy 2004;77:203–16.
[63] Veerachary M, Yadaiah N. ANN based peak power tracking for PV supplied DC motors. Sol Energy 2000;69:343–50.
[64] Mazouz N, Midoun A. Control of a DC/DC converter by fuzzy controller for a solar pumping system. Int J Electr Power Energy Syst 2011;33:1623–30.
[65] Kassem AM. MPPT control design and performance improvements of a PV generator powered DC motor-pump system based on artificial neural networks. Int J Electr Power Energy Syst 2012;43:90–8.
[66] Mozaffari Niapour SAK, Danyali S, Sharifian MBB, Feyzi MR. Brushless DC motor drives supplied by PV power system based on Z-source inverter and FL-IC MPPT controller. Energy Convers Manag 2011;52:3043–59.
[67] Zaki AM, Eskander MN. Matching of photovolatic motor-pump systems for maximum efficiency operation. Renew Energy 1996;7:279–88.
[68] Eskander MN, Zaki AM. A maximum efficiency-photovoltaic-induction motor pump system. Renew Energy 1997;10:53–60.
[69] Akbaba M, Qamber I, Kamal A. Matching of separately excited dc motors to photovoltaic generators for maximum power output. Sol Energy 1998;63:375–85.
[70] Chary MV. Steady-state and dynamic performance analysis of PV supplied DC motors fed from intermediate power converter. Sol Energy Mater Sol Cells 2000;61:365–81.
[71] Terki A, Moussi A, Betka A, Terki N. An improved efficiency of fuzzy logic control of PMBLDC for PV pumping system. Appl Math Model 2012;36:934–44.
[72] Akbaba M, Akbaba MC. Dynamic performance of a photovoltaic-boost converter powered DC motor-pump system. In: IEEE International Conference in Electric Machines and Drives IEMDC 2001; 2001. p. 356–61.
[73] Dubey M, Sharma S, Saxena R. Solar PV stand-alone water pumping system employing PMSM drive. In: IEEE Students' Conference in Electrical, Electronics and Computer Science (SCEECS); 2014. p. 1–6.

CHAPTER 3

Modeling of photovoltaic water pumping systems

Contents

3.1 Photovoltaic water pumping system components

In general, a typical photovoltaic water pumping system consists of four main parts: a PV array, a DC-DC converter, a permanent magnet DC motor coupled with a centrifugal pump, and a tank that serves as a hydraulic storage device, as shown in Fig. 3.1. The storage tank supports the system when the PV array is unable to power the pump at night or when the volume of pumped water is insufficient to cover the demand for water. In this chapter, the mathematical models of photovoltaic water pumping system components are discussed in the following sections.

3.2 Photovoltaic array model

A solar cell can be represented as a double-diode model, as shown by the circuit in Fig. 3.2. The output current of a solar cell can be expressed by

$$I_c = I_{Ph} - I_{o1}\left[\exp\left(\frac{V_c + I_c R_s}{V_{t1}}\right) - 1\right] - I_{o2}\left[\exp\left(\frac{V_c + I_c R_s}{V_{t2}}\right) - 1\right] - \frac{V_c + I_c R_s}{R_p}, \tag{3.1}$$

where I_c and V_c are the output current (A) and voltage (V) of the solar cell, respectively; I_{Ph} is the photocurrent; I_{o1} and I_{o2} are the diode saturation currents of the first and second diodes, respectively (A); R_s is the series resistance (Ω);

https://doi.org/10.1016/B978-0-12-821231-8.00003-2

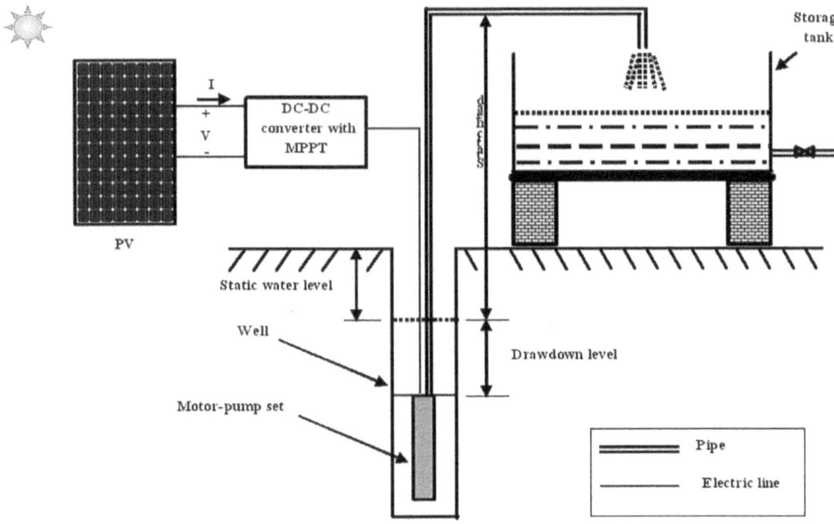

Fig. 3.1 PV water pumping components.

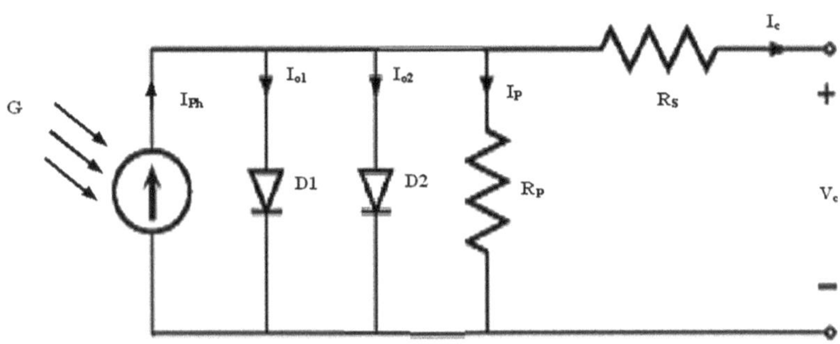

Fig. 3.2 Solar cell double diode model.

R_p is the shunt resistance (Ω); and V_{t1} and V_{t2} are the diode thermal voltages that can be given by

$$V_{t1} = \frac{a_1 kBT_C}{q}, \tag{3.2}$$

$$V_{t2} = \frac{a_2 kBT_C}{q}, \tag{3.3}$$

where q is the electron charge (1.60217646e-19C), KB is Boltzmann's constant (1.3806503e-23 J/K), T_C is the cell temperature (K), and a_1 and a_2 are the diode ideality factors that represent the components of the diffusion and recombination currents, respectively.

The PV array consists of the N_s and N_p modules, which are connected in series and parallel, respectively, to constitute the power demand by the load. Thus, the output current of a PV array can be expressed by

$$I_a = N_p I_{Ph} - N_p I_{o1}\left[\exp\left(\frac{1}{V_{t1}}\left(\frac{V_a}{N_s}+\frac{I_a R_s}{N_p}\right)\right)-1\right] - N_p I_{o2}\left[\exp\left(\frac{1}{V_{t2}}\left(\frac{V_a}{N_s}+\frac{I_a R_s}{N_p}\right)\right)-1\right] - \frac{N_p}{R_p}\left(\frac{V_a}{N_s}+\frac{I_a R_s}{N_p}\right), \quad (3.4)$$

where V_a and I_a are the output voltage (V) and current (A) of the PV array, respectively.

The efficiency of the PV array can be computed as

$$\zeta_{PV} = \frac{V_a I_a}{A G_T}, \quad (3.5)$$

where A is the area of the PV array (m^2) and G_T is the hourly solar radiation on the PV array surface (W/m^2).

In general, the most important issue to accurately model a PV array is to find optimal values of the seven parameters (I_{Ph}, I_{o1}, I_{o2}, R_s, R_p, a_1,and a_2) in Eq. (3.1). These parameters can be obtained by an experiment that extracts a number of the I-V curve. Based on the extracted I-V curve data, optimization can be applied to these data whereas the optimum value of these seven parameters can be obtained. However, it is also worth mentioning that the seven parameter values can also be obtained using the I-V curves tested at standard testing conditions that are provided in the data sheets. Here, the value of these parameters may not imply any site characteristic that may affect the performance of the PV panel.

On the other hand, there are other methods that can be used for modeling a PV panel/array following the fact that the meteorological data have a big effect on the performance of the PV system. Changing solar radiation affects the output current produced by a PV module/array proportionally. On the other hand, the ambient temperature affects the output voltage of the cells proportionally. However, considering the average ambient temperature and load line location in the designing phase, the output voltage of a PV array/module does not vary significantly during a day, as the ambient temperature has a low swinging throughout the day. Meanwhile, the solar radiation is always changing and consequently the output current of a PV module/array is expected to fluctuate throughout the day time. In other words, the output power of a PV system mainly depends on the output current of the PV module/array.

A typical PV system usually consists of a PV array, power conditioners such as maximum power point trackers and charge controllers, a battery, an inverter, and a load. For the PV array model, various mathematical models have been introduced to describe it. The general working concept of the PV system is that the incident radiation of the sun on the PV array is collected and converted to a DC current. This DC current powers the load after passing through a controller and an inverter. The output power of a PV array strongly depends on the solar radiation and the ambient temperature. Hence, the equation below describes the output current of a PV array.

$$I_{PV}(t) = \frac{\left[P_m\left(\frac{G_T(t)}{G_{reference}}\right) - \alpha_T\left(T_c(t) - T_{reference}\right)\right] \times \eta_{inv} \times \eta_{wire}}{V_{PV}(t)} \tag{3.6}$$

where G_t is the correlated solar radiation in (W/m^2), $G_{reference}$ is the solar radiation at reference conditions in (W/m^2), α_T is the temperature coefficient of the PV module power that is given by the manufacturer, $T_{reference}$ is the cell temperature at reference conditions, η_{inv} and η_{wire} are the efficiencies of the inverter and the wires, respectively, and V_{PV} is the array voltage. T_c is the cell temperature and it can be calculated by the following equation:

$$T_c(t) = T_{amb}(t) + \left(\left(\frac{\text{NOCT} - 20}{800}\right) \times G_t(t)\right) \tag{3.7}$$

where T_{amb} is the ambient air temperature in °C and NOCT is the normal operating cell temperature in °C. The NOCT represents the cell temperature of a PV module when the ambient temperature is 20°C, solar radiation is $800\,W/m^2$, and the wind speed is 1 m/s.

In addition, regression analysis can be considered one of the most famous techniques used for analyzing multifactor data. Regression analysis is a statistical process to predict and express the relationships between the variables of interest (dependent variable and independent variables). The simplest regression model is represented by a simple linear regression model, which is a model with a single explanatory variable (x) that has a relationship with the response (y) in the straight line, as illustrated below:

$$y = \beta_o + \beta_1 x \tag{3.8}$$

where β_o is the intercept and β_1 is the slope of the line.

On the other hand, another form of regression analysis is the multiple regression model. This model considers more than one independent variable. In other words, multiple regressions simultaneously consider the influence of multiple explanatory variables on a response variable. The basic model for linear multiple regression is:

$$y = \beta_o + \beta_1 x_1 + \beta_2 x_2 \tag{3.9}$$

where β_o the intercept and (β_1, β_2) are the regression coefficients.

The multiple regression model can be used to predict the output current (I_{PV}) of a PV array using meteorological parameters such as ambient temperature (T) and solar radiation (G).

As for the efficiency calculation of the power conditioning unit, the efficiency is described in terms of its input power and its rated power. The efficiency curve can be described by a power function as follows:

$$\left\{\begin{array}{ll} \eta = C1\left(\frac{P_{PV}}{P_{PCU_R}}\right)^{C2} + C3 & \frac{P_{PV}}{P_{PCU_R}} > 0 \\ \eta = 0 & \frac{P_{PV}}{P_{PCU_R}} = 0 \end{array}\right\} \tag{3.10}$$

where P_{PV} is the PV module output power and P_{PCUR} is the power conditioning unit's rated power while $C1$–$C3$ are the model coefficients.

3.3 Pump-motor model

A brushless permanent magnet DC (BPMDC) motor is used in this study to overcome the obstacles that beset the conventional DC motor for low power systems. According to the output voltage and current of the PV array at MPP, the back electromotive force (EMF) of the DC motor can be calculated by

$$V = IR_a + E_b, \tag{3.11}$$

where V and I are the armature voltage (V) and current (A) of the DC motor, respectively; E_b is the motor back EMF voltage (V); and R_a is the armature resistance (Ω). The electromechanical torque of the DC motor is represented as follows:

$$T_m = K_T I, \tag{3.12}$$

where T_m is the electromechanical torque of the DC motor (Nm) and K_T is the motor torque constant (Nm/A). Neglecting the rotational losses, the mechanical output power of the DC motor (P_{mo}) can be represented as in Eq. (3.12), and its efficiency (ζ_M) can be calculated as below

$$P_{mo} = I^* E_b, \tag{3.13}$$

$$\zeta_M = \frac{P_{mo}}{I^*V}, \tag{3.14}$$

When the friction losses are neglected, the electromechanical torque of the motor is equal to the torque required to pump a certain quantity of water (i.e., the torque of the pump). The torque of the pump can be represented as a function of the rotational speed of the pump:

$$T_P = K_P{}^*\omega^2, \tag{3.15}$$

where ω is the rotational speed of the DC motor (rad/s) and K_P is a constant computed based on the shaft and impeller dimensions of the pump, as described below:

$$K_P = 2\pi\rho b_1 R_1^2 \tan(\beta_1)\left(R_2^2 - \frac{R_1^2 b_1 \tan(\beta_1)}{b_2 \tan(\beta_2)}\right), \tag{3.16}$$

where ρ is the water density (Kg/m^3); R_1 and R_2 are the impeller radius at the impeller inlet and outlet, respectively (mm); b_1 and b_2 are the heights of the impeller blade at the impeller inlet and outlet, respectively (mm); and β_1 and β_2 are the inclination angles of the impeller blade at the impeller inlet and outlet, respectively (degree). The rotational speed can be calculated as

$$\omega = \sqrt{\frac{K_T}{K_p} I}. \tag{3.17}$$

Neglecting the friction losses, the input power of the pump (P_{pi}) is equal to the mechanical output power of the motor. The output power and efficiency of the pump (P_{po}) can be computed by Eqs. (3.18) and (3.19), respectively:

$$P_{po} = T_p{}^*\omega, \tag{3.18}$$

$$\zeta_P = \frac{P_{po}}{P_{pi}}. \tag{3.19}$$

The subsystem efficiency (ζ_{sub}) and overall efficiency of the PVPS (ζ_{sys}) are calculated as follows:

$$\zeta_{sub} = \zeta_M{}^*\zeta_P, \tag{3.20}$$

$$\zeta_{sys} = \zeta_{PV}\zeta_M\zeta_P. \tag{3.21}$$

The total head (H) at any time of the operation of the photovoltaic water pumping system can be represented as follows:

$$H = H_s + H_{dd} + H_D + H_d, \tag{3.22}$$

where H_s is the static head and is equal to the difference between the surface of the water and the discharge point, H_{dd} is the drawdown water level, and H_D and H_d are the equivalent heads due to friction losses in the pipeline and fitting components, respectively. The pipeline friction losses (H_D) are a dynamic head and are computed based on the Darcy-Weisbach formula:

$$H_D = \delta \frac{Lv^2}{2gd}, \tag{3.23}$$

where L is the length of the pipeline (30 m), d is the internal diameter of the pipeline (0.0508 m), g is the acceleration due to gravity (m/s^2), δ is the pipeline friction coefficient dependent on the Rynold's number (i.e., 0.2461), and v is the average speed of the water (m/s) that is related to the water flow rate and the cross-sectional area of the pipeline as described below:

$$v = \frac{4Q}{\pi d^2}, \tag{3.24}$$

where Q is the water flow rate (m^3/s). The friction losses due to the fitting components, such as valves, junctions, pipe entries, and elbows, can be computed with the following formula:

$$H_d = \beta \frac{v^2}{2g}, \tag{3.25}$$

3.4 Storage tank model

As for the storage tank model, it should reflect the status of the water level in the tank considering the charged and discharged water, The current resident water in the storage tank, the deficit water volume, the excess water volume, and the state of charge of the storage tank can be evaluated every hour over a year as follows:

$$C_{res}(t) = \begin{cases} C_{res}(t-1) + Q(t) - D & \text{if } (C_{res}(t-1) + Q(t)) > D \\ 0 & \text{otherwise} \end{cases} \tag{3.26}$$

$$Q_d(t) = \begin{cases} |C_{res}(t-1) + Q(t) - D| & \text{if } (C_{res}(t-1) + Q(t) - D) < 0 \\ 0 & \text{otherwise} \end{cases} \tag{3.27}$$

$$Q_e(t) = \begin{cases} C_{res}(t-1) + Q(t) - D - C_n & \text{if } (C_{res}(t-1) + Q(t) - D) > C_n \\ 0 & \text{otherwise} \end{cases} \tag{3.28}$$

$$SOC(t) = \frac{C_{res}(t)}{C_n}, \tag{3.29}$$

where $C_{res}(t)$ is the current resident water in the storage tank (m^3), D is the hourly water demand (m^3/h), Q_d is the hourly water deficit (m^3), Q_e is the hourly excess water (m^3), $SOC(t)$ is the hourly current state of charge of the storage tank, and C_n is the maximum capacity of the storage tank (m^3).

CHAPTER 4

Intuitive and numerical sizing methods for photovoltaic water pumping systems

Contents

4.1 Intuitive method for sizing photovoltaic water pumping systems

In general, the first requirement for designing a photovoltaic water pumping system is an estimate of the water needed and the amount of water that can be supplied by the source (flow rate). If the water needs vary throughout the year, a monthly profile should be drawn and matched to a monthly profile of the production capability of the water source. It is important also to know the worst case conditions, so data on production and demand for the driest months of the year should be available or estimated. Here, if the capability of the water source is limited, the designer must take action. One thing that can be done is to improve the water source or develop other sources. Using a smaller pump is another option, but the availability of different sized pumps is limited. Another method is to incorporate batteries into the system and distribute the pumping time over a longer period. This is one of two reasons to use batteries in a water pumping system. The other is if the pumping time needs to be controlled in order to pump at a high flow rate for a short time. An example might be a residential system with storage tanks when you want

Photovoltaic Water Pumping Systems
https://doi.org/10.1016/B978-0-12-821231-8.00004-4

to pump all the water for the household during times when other loads are not operating. Although using batteries in a system will maximize the pump efficiency because of the steady operating conditions presented to the pump and motor, most water pumping systems do not contain batteries. The reason behind that is storing water is usually less expensive than storing electricity. If a tank is available, the system can pump all day and the water is stored for later use. Gravity feed or a small pressure pump can then be used to deliver the water to the user.

Another variable that must be specified when designing a photovoltaic water pumping system is the pumping time factor. This factor is related to the number of daily peak sun hours. If a direct-drive centrifugal pump is used, the pumping time factor will equal 1.0. In other words, the pump will operate with varying efficiency through all daylight hours, but that is equivalent to operating at the rated efficiency during peak sun hours. If batteries are used, the pumping time factor would be equal to the hours of scheduled operation divided by the number of peak sun hours. If a linear current booster or peak power tracking controller is used between the array and the pump in a direct drive system, the pumping time factor should be 1.2. This takes into account the improvement in pump performance that these devices achieve.

The pump size, operating time, and total power demand can be calculated if the efficiency of the pump and the depth of the water are known. The efficiency of specific pumps depends on the pump type and operating conditions. For centrifugal pumps, the efficiency is a function of head, flow, and solar insulation, all of which will vary throughout the day. Under some conditions, the average daily efficiency, called the wire-to-water efficiency, can be as little as one-third the peak pump efficiency. In contrast, the efficiency of a displacement pump changes little with changing solar conditions. Some typical wire-to-water efficiencies are given in Table 4.1.

Table 4.1 Measurement of wire-to-wire efficiency.

Head (m)	Pump's type	Wire-to-wire efficiency (%)
0–5	Centrifugal	12–25
6–20	Centrifugal with jet	10–20
	Submersible	20–30
21–100	Submersible	30–40
	Jack pump	30–45
>100	Jack pump	35–50

Many pumping systems use PV arrays mounted on one-axis trackers. Tracking the array not only increases the hours of operation (peak sun hours) but also provides a more consistent operating point (voltage and current) for the pump motor. Therefore, tracking is recommended.

The two most important factors in the operation of a PV pump are the availability of sufficient solar radiation to enable the pump to start (until the solar radiation reaches the threshold level), and the nonlinear relationship between the pumping rate and solar radiation. The threshold level of a PV pump depends on the system components. Fig. 4.1 illustrates typical PV pump components. It consists of a PV array, an inverter, the motor-pump subsystem, and the water tank.

These factors are water source capacity, water volume required per day, solar insulation availability, pumping time, static water level, drawdown level, discharge head, pipe size friction, and pumping subsystem efficiency. The most important is total dynamic head (TDH), which is the sum of the static head, the drawdown, and the equivalent head caused by friction losses in the pipe. TDH is expressed in feet or meters and is dependent on the flow rate. Fig. 4.2 shows details about TDH components.

PV pump components have to be selected carefully for a proper matching of the system. Unlike conventional pumping systems, PV pumps have to be designed and installed properly to be competitive with other pumping technologies. Each component of a PV pump has intrinsic characteristics affecting the overall operating conditions. Therefore, it is desirable that

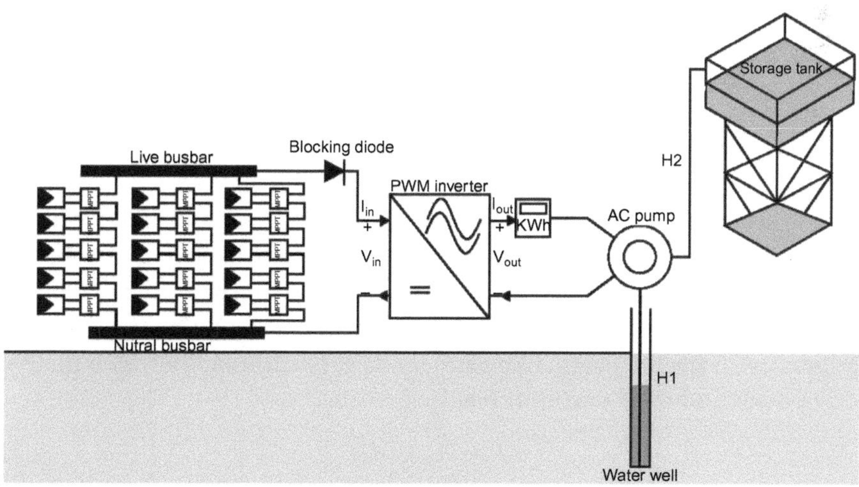

Fig. 4.1 Diagram of typical PV pump components.

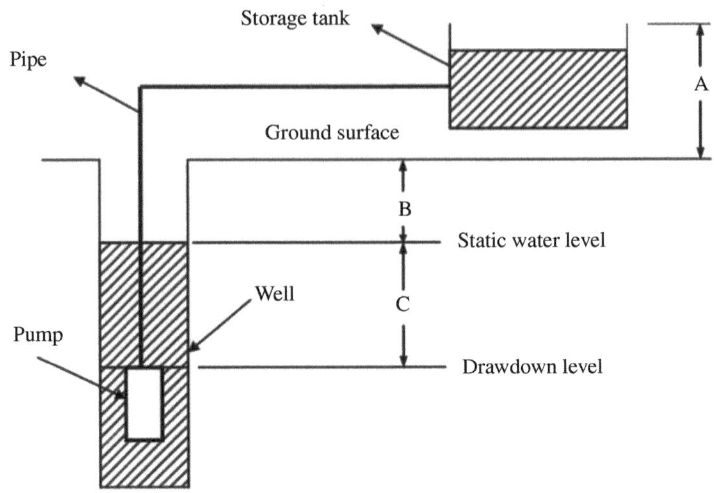

Fig. 4.2 Water pumping system terms. These factors are water source capacity, water volume required per day, and solar insulation.

the intercept of all respective component characteristics follows the maximum power line of the PV array generator. Depending on this internal matching, the efficiency of the overall system and the related performances will meet an acceptable range. System design, particularly the PV array capacity, should be reviewed to ensure that sufficient energy is produced to start the motor as early in the day as possible. In principle, modeling components individually and combining them into a single system can optimize PV pumps. This approach could be the most accurate method to maximize the efficiency of each component and, ultimately, the overall system efficiency. This method requires an understanding of each component. Detailed information, such as the ideality factor, the operating principles of diodes, the shunt and series resistance, etc., is not easily ascertained from catalogues. To use this approach, one must know the basic functions of diodes, inverters, motors, and pumps. Another simple and fairly accurate design method is to create a clear mathematical relationship between the solar radiation energy, the PV array power, and the required hydraulic energy to fulfill the water demand. This method can easily be used by field technicians or by end users. The mathematical relationship between the PV array power and solar radiation energy is:

$$P = A_{pv} G_r \eta_r \tag{4.1}$$

where P is the PV array power (Watt), η_r is the efficiency of the PV array at a reference temperature ($T_r = 25°C$), G_r is the solar radiation at a reference temperature ($G_r = 1000\,W/m^2$), and A_{pv} is the effective area of the PV array, in m^2 ($A_{pv} = n_p n_s A$, where A is the area of a single module, and n_s is the number of the group of PV cells connected in series, each containing n_p strings in parallel).

The effective PV array area is calculated from the relationships of the daily energy output E_e and the daily hydraulic energy E_h (both in kWh):

$$E_e = A_{pv} G_T \eta_{pv} \tag{4.2}$$

and

$$E_h = \rho g h V \tag{4.3}$$

where $\boldsymbol{\eta_{pv}}$ is the efficiency of the PV array under operating conditions, G_T is the daily solar radiation on the PV array surface (kWh/m^2), V is the daily amount of water required (m^3), h is the total pumping head (m), $\boldsymbol{\eta_s}$ is the subsystem efficiency, $\boldsymbol{\rho}$ is the density of water, and g is the acceleration due to gravity. The efficiency of the PV array is determined from:

$$\eta_{pv} = f_m[1 - \alpha(T_c - T_r)]\boldsymbol{\eta_r} \tag{4.4}$$

where f_m is the matching factor, that is, the ratio of the power output of the PV array under operating conditions to its power output at the maximum power point. The generally accepted value for designing a PV system is $f_m \approx 0.90$. The value α is the cell temperature coefficient and is from 0.2% to 0.6%/C (0.004 to 0.005/C for Si), and T_c is the daily average cell temperature (in C). Cell temperature is dependent on solar radiation, ambient temperature, and wind speed.

The simplified cell temperature model that was commonly adopted by many researchers is the relation that includes solar radiation and ambient temperature. The relation is of the form:

$$T_c = T_a + \frac{G_T}{800}(NOCT - 20) \tag{4.5}$$

where T_a is the hourly ambient temperature and $NOCT$ is the module junction temperature under normal operating temperature ($G = 800\,W/m^2$, $T_a =$ 20C and at 1 m/s wind speed). For a wind speed over 1 m/s, the cell (module) temperature will be lower, and the cell temperature decreases as the wind speed increases.

Once the efficiency of the PV array is determined, the PV array area can be calculated as below,

$$A_{pv} = \frac{\rho g h V}{G_T \eta_{pv} \eta_s} \tag{4.6}$$

The overall efficiency of a PV pump can be determined from the hydraulic energy and from the solar radiation energy input as below,

$$\eta_o = \frac{P_h}{P_{in}} = \frac{\rho g h V}{A_{pv} G_T} \tag{4.7}$$

A simple nomogram can be developed using the aforementioned relationships at various ambient air temperatures. The nomogram shown in Fig. 4.3 was developed for ambient air temperature T_a of 25°C.

This nomogram can be used either to determine the size of a PV array required for the desired pumping head and water demand, or conversely to estimate the daily amount of water production for a given PV array size and pumping head. If the hydraulic energy varies from month to month as a result of variation in water levels or water demand, the nomogram should be used for each month of the year with the corresponding solar radiation and hydraulic energy. The month with the worst-case combination of solar radiation energy and water demand is usually the design month. To use this nomogram, first determine the water demand from the size of the population to be served and the total pumping head. Then draw a line counterclockwise on the nomogram, using the appropriate values of the subsystem's efficiency and the average daily solar radiation, to get the required array size. Alternately, start clockwise from a known PV array size to obtain the amount of water produced for a given total pumping head.

4.2 Design examples on sizing photovoltaic water pumping systems using the intuitive method for livestock in Nablus city

Nablus is a city in Palestine that is surrounded by many villages where a lot of people shepherd cattle. Most of the cattle farms (mainly cow farms) are located in areas that are quite far from a grid connection. Thus, rural electric cooperatives in Nablus are beginning to supply PV power to ranchers who need to pump water for livestock. One of these ranchers requires about 500 gal of water per day for 40 cows. The pasture where the well is located is used in wintertime only. There is a 2500 gal tank near the well that provides water (gravity fed) to smaller water tanks distributed throughout the pasture. The load is critical as the cattle cannot go without water more than 1 day in winter. In this farm, there is a small pump that uses a 12 V DC and is designed to pump more than 500 gal per day (2000 L per day) from a level of 20 m. The water level in this well is only 10 m with a maximum drawdown

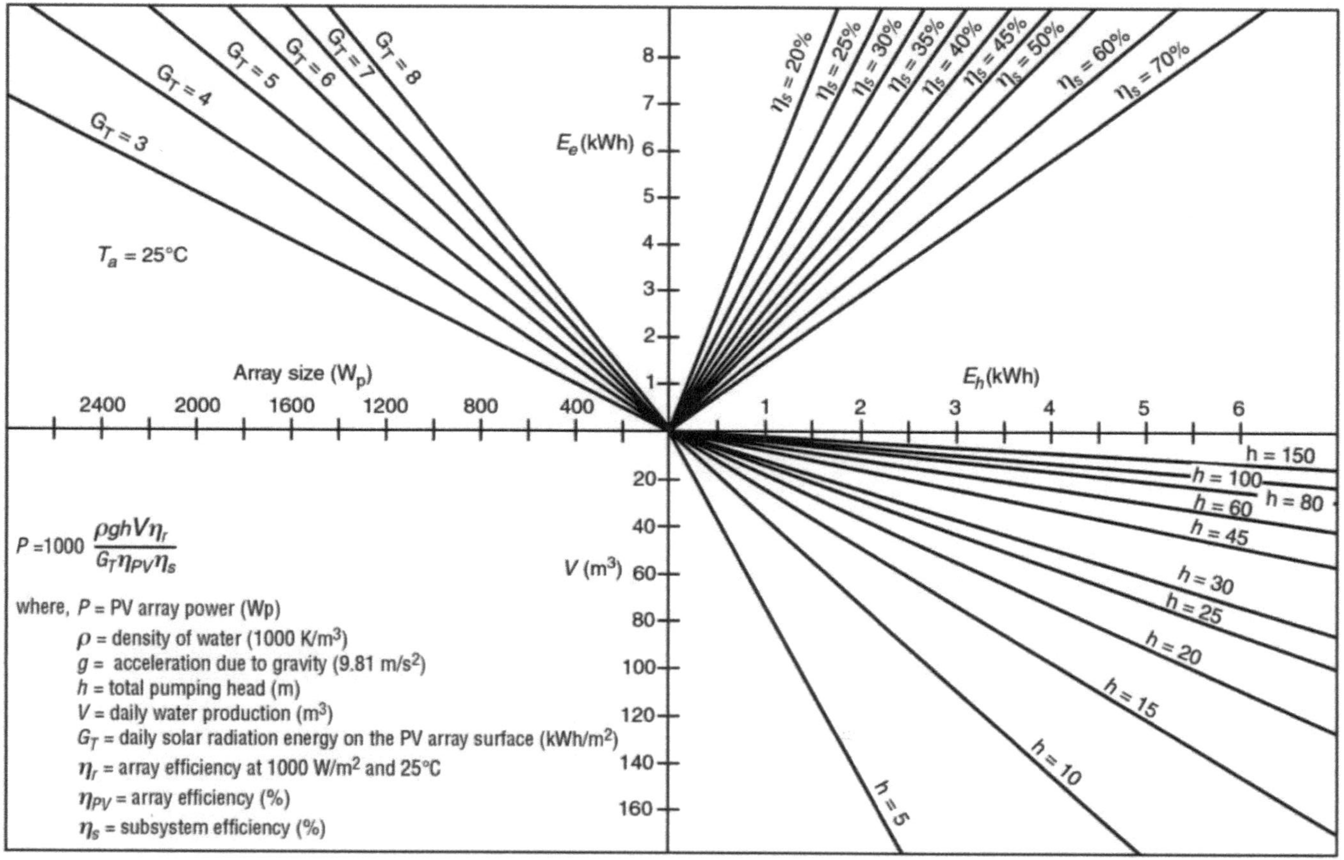

Fig. 4.3 Nomogram used to estimate the size of a PV array and daily water production.

of 6 m (dynamic head (m): 17). Finally, there is a float switch in the main watering tank that controls the operation of the pump.

On the other hand, the average solar irradiation is 5.5 kWh/m^2.day where the minimum solar irradiation (2.79 KWh/m^2.day) is in January and December. The maximum value of solar irradiation is usually in June, which is 8.74 kWh/m^2.day. The average ambient temperature in Nablus is in the range of 14–26°C.

Here, it is required to design an appropriate PV system so as to power the motor pump set available at the farm where 500 gal of water can be pumped every day of the year, especially in winter (water required/day: 2000 L, October to March).

To solve this problem we have to estimate the pumping rate (L/h), which can be given by:

$$\text{Pumping}_{\text{Rate}}\left(\frac{\text{L}}{\text{hour}}\right) = \frac{\text{Water requierd per day}}{\text{Pumping time factor} \times \text{PSH}} \tag{4.8}$$

Using (4.8), assuming that the pumping time factor is (1.2), then the system must pump 243 L/h.

The required energy (Wh/day) can be calculated then by:

$$\text{Hydraulic energy}\left(\frac{\text{Wh}}{\text{day}}\right) = \frac{\text{Water required per day} \times \text{total dynamic head}}{367\ (\text{conversion factor})} \tag{4.9}$$

Using (4.9), the hydraulic energy is 93 Wh/day. Then the required PV array energy is given by the following:

$$E_{array} = \frac{\text{hydraulic energy}}{0.25\ (\text{pump system efficiency})} \tag{4.10}$$

The energy of the PV module array is 372 Wh/day.

Suppose that the system's nominal voltage is 12 V. Then the ampere hour load is 31 Ah/day.

As for the design current and based on the PSH for Palestine, the worst months are January and December, which are both 2.97 KWh/m^2-day. Therefore, the design current is the load current divided by the worst PSH, which is 10.5 A.

As for PV array size, suppose that the PV module's rated current and voltage are 3 A and 15.9 V, respectively. Then the number of PV modules in parallel is the design current divided by the PV module's rated current, which is equal to three PV modules. One PV module is needed in series because the system's nominal voltage is 12 V.

Now, based on the calculated size of the PV system, the pumping rate (L/day) is given by:

$$\text{Pumped water}\left(\frac{\text{L}}{\text{day}}\right)=\frac{N_p I_m V_{nominal}\eta_{system}\times\text{converion factor}\times PSH}{TDH} \tag{4.11}$$

where N_p is the number of modules in parallel, I_m is the module current, $V_{nominal}$ is the system nominal voltage, PSH is the peak sunshine, and TDH is the total dynamic head. Using (4.11), the pumped water per day is 3206 L/day or 486 L/h.

4.3 Design example for photovoltaic water pumping systems using the intuitive method for the oasis in Mauritania

"Oasis" is a Greek word that means a fruitful zone in a desert and arid environment. The oasis has its own microclimate that causes the "oasis effect," which means that the ambient temperature in the desert area is cooler than the surrounding desert. Oases have the ability to adapt to harsh environments; therefore, they are considered instruments to fight the desertification of arid regions.

Saharan and Arabian deserts have many hydrogeological basins that are distributed through many countries, including Mauritania. Mauritania is an Arab country located in Northwest Africa. It is located between 15°45′ and 27°25′ north and 4°10′ and 16°50′ west. The country has two major climate zones: the Saharan Zone in the north and the Sahelian Zone in the south. The oases areas are distributed over four regions in Mauritania: Adrar, Tagant, Assaba, and Hodh. The date palm is the essential part of the oases and has a unique heritage in the country's culture. It is associated with Mauritanian traditions and roots. Mauritanian oases have 2.4 million date palms distributed over the four regions. Annually, the 65% yield of date palm trees can reach 20,000 MT of fruit on average. Moreover, Mauritania accounts for 10.7% of date palm production in the Maghreb region.

There are some natural and technical problems regarding the date palm sector; these can be classified as nature and technical problems. The natural problems are represented by a difficult desertic hot climate with sand dunes, narrow valleys, and frequent drought. Meanwhile, there are technical problems related to traditional and inefficient cultivation practices such as flood irrigation through traditional channels. Date farmers, in general, have lack of knowledge and awareness of modern palm farming techniques as well as a lack of management of the available water. Mauritania suffered from drought since the 1970s, which has made farmers use many pumping systems, from manual withdrawal pumping to diesel motor pumps, to exploit the groundwater. At that time, the capacity of the available pumps was higher than needed, which can lead to well depletion. In fact, accessibility to water is limited to just 20% of the global cultivated land and less than 5% in sub-Saharan Africa.

The Adrar Region in Mauritania (approximately 223,000 km^2) has 57 oases and the Tagant Region (approximately 97,000 km^2) has 60 oases. Both regions depend majorly on raising livestock and date cultivation as sources of livelihood.

Underground water sources (aquifers) supply most of the oases with water. In many cases, a natural spring gets the underground water to the surface. Meanwhile, in some other cases, man-made wells are constructed. There are three types of groundwater found in the oasis regions: groundwater along wadis (Arabic name for a valley), groundwater under sand dunes, and fissure water along fracture zones in basement rocks. Groundwater along wadis is more widespread than other types. The recharge of the groundwater aquifer comes from the percolation of intensive rain. In the Adrar region, about 5% of the wells are pumped by motor pumps and the deep wells are provided with submersible pumps that are powered with gasoline or diesel generators.

The regions have rainy seasons with a high temperature from July to October, dry seasons with low temperature from November to March, and dry seasons with a high temperature from April to June. Regarding air temperature, the average is about 28.4°C. The maximum air temperature reaches 34.9°C while the minimum temperature is 24.4°C in Adrar. The daily average of sunshine in the Adrar region reaches 8.3 h. The average monthly wind speed can reach 3 m/s in Atar and the dominant direction is from north to west. Therefore, solar energy-based systems are recommended as compared to wind energy-based systems.

The water wells in the adapted region have a circular shape with an inner diameter of 0.8–2 m. The water wells in Adrar have depths between 8 and 16 m. The flow of the water wells is 2.2 m^3/h for Adrar and 1.8 m^3/h for Tagant. Based on these results, the potential of the water well over a day would be an average of around 48 m^3/day.

On the other hand, Mauritania has a massive area with a high exposure of solar radiation that makes it a perfect destination for solar energy projects. The average solar radiation is 4.9–6.5 $kWh/m^2/day$ on a horizontal surface.

Thus, this example aims to design a photovoltaic water pumping system for irrigation purposes for an oasis in Mauritania. The design will be done for

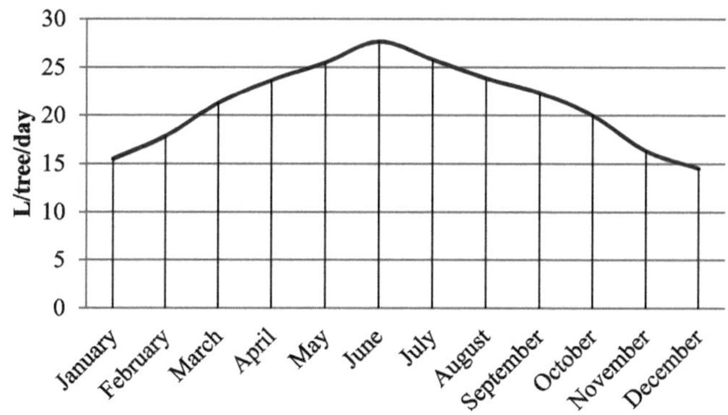

Fig. 4.4 Daily water requirement per tree in the Adrar region.

the Tawaz oasis, which is one of the biggest oases in date palm production in Atar city in the Adrar region in Mauritania. The Tawaz oasis basically consists of three main regions: Agadir in the north, and Batha' and Lmeileh in the south. The northern region has about 3000 trees while the southern region has about 17,000 trees.

Fig. 4.4 shows the water requirement for date palm trees in the Adrar region.

As for the water source, there are eight water wells in the Tawaz oasis. Table 4.2 summarizes the specifications of the wells in the selected region.

Table 4.3 shows the maximum number of trees that each well should provide.

The northern region has three wells; therefore, the total water requirement will be divided over the three wells, which equals 107.84 m^3/day. Thus, the flow rate of the pumps equals 19 m^3/h. On the other hand, the southern region has five wells, and each well has a different capacity. Thus, the total water requirement will be distributed unevenly in which each well can provide 80% of its capacity. Table 4.4 summarizes the water requirement that each well should provide and the flow rate for each pump:

After the water requirement and the well specification are defined, the sun peak hours of the adopted region should also be defined. The Tawaz Oasis is located in Atar city while Atar has an average solar radiation or peak sun hours of 5.67 kWh/m^2/day.

One other thing that is required for design is the pump's total dynamic head (TDH). The slope of the topography has to be taken into account as well. In the case study design, C-PVC has been selected as the pipe material ($\varepsilon = 1.524\,\mu m$) due to its low roughness value, high temperature strength, light weight, toughness, flexibility, ability to withstand temperatures up

Table 4.2 Well specifications in Tawaz oasis.

Location	Elevation (m)	Drawdown level (m)	Static water level (m)	Well's capacity (m^3/day)	Well's capacity (m^3/ h)	Well's name
12° 52′ 06′W 20° 46′ 24′N	309	16.72	18	124.8	10.4	Agadir well 1
12° 52′ 12′W 20° 46′ 23′N	309	12.13	14	148.8	12.4	Agadir well 2
12° 52′ 27′W 20° 46′ 22′N	311	10.2	12	130.8	10.9	Agadir well 3
12° 53′ 14′W 20° 42′ 03′N	285	2.74	27	185.16	15.43	Batha' well 1
12° 53′ 14′W 20° 42′ 05′N	288	3.4	34	205.56	17.13	Batha' well 2
12° 53′ 15′W 20° 42′ 06′N	288	2.84	28	177.6	14.8	Batha' well 3
12° 54′ 00′W 20° 41′ 13′N	286	1.78	77	147.6	12.3	Lemeileh well 1
12° 53′ 56′W 20° 40′ 57′N	284	1.12	93	74.4	6.2	Lemeileh well 2
–		–	–	1194.72	99.56	Total

Table 4.3 Maximum number of trees each well can irrigate.

Max number of trees	Well's name
3068	Agadir well 1
3658	Agadir well 2
3216	Agadir well 3
4552	Batha' well 1
5053	Batha' well 2
4366	Batha' well 3
3629	Lmeileh well 1
1829	Lemeileh well 2

Table 4.4 Pump flow rate for each well.

Pump flow rate (m^3/h)	Water requirement/ well (m^3/day)	Well's name
26.5	150	Batha' well 1
30	168	Batha' well 2
25	142	Batha' well 3
21	118	Lemeileh well 1
10	55	Lemeileh well 2

to 93 °C, excellent chemical resistance, and the low costs of its maintenance and installation.

Selection of the pump and power requirement: The pumps are selected using the pump's performance curves, which represent the intersection of the flow rate and the TDH that gives the appropriate power that can run the pump. Pump selection can be done using Fig. 4.5.

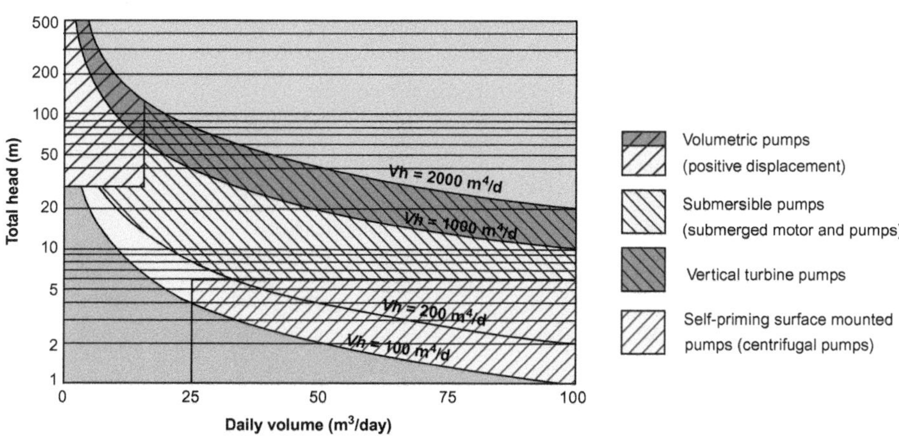

Fig. 4.5 Selection of the pumps based on TDH and flow rate.

As for PV modules and inverter selection, thin-film PV modules are used in this research. The specifications of the PV modules under NOCT are summarized in Table 4.5.

Because the motors of the submersible pumps are three-phase induction motors, the inverters will be used to convert the DC of the PV modules into AC to run the motors. Renewable solar pumping inverters have been used in the design example. The efficiency of the inverter has been taken as 98%. Table 4.6 describes the specifications of the selected inverters.

In general, the first thing to design in a PV pumping system is the storage tank. According to the gathered information, June has the peak water demand throughout the year (27.66 L/tree). Thus, the design has been done based on this month (the worst month method). Based on that, the daily total water requirement for the northern region equals 97.62 m^3/day, taking into account an 85% efficiency of drip irrigation. Meanwhile, for the southern region, this is assumed to be equal to 553.2 m^3/day. The total capacity of the northern region's wells and the southern region are 404.4 and 790.32 m^3/day, respectively. Regarding the water storage tank, the elevation head of the reservoir's stand is found to be 2 m, as the drip irrigation requires an elevation head between 1 and 2 m. A fiberglass storage tank can come as a standard-made unit. Based on the proposed design, each tank can irrigate 100 date palm trees distributed over approximately 1 ha. The size of the tank has been calculated based on the palm tree requirement in June

Table 4.5 PV module specifications.

Value	Specifications
320.9 W	Nominal power at NOCT P_{max}
169.8 V	Voltage at P_{max}
1.89 A	Current at P_{max}
206.6 V	Open-circuit voltage V_{oc}
2.05 A	Short-circuit current I_{sc}

Table 4.6 Inverter specifications.

Value	Specifications
3 × 380–415 V	Voltage
400 V	Min MPPT voltage
800 V	Max MPPT voltage
5 Hz	Min frequency
60 Hz	Max frequency

with a 25% safety factor. Thus, the tank has an approximate capacity of 4 m^3 with a 1.2 m diameter and a 3 m height.

In the proposed design, the flow rates of the pumps are between 10 and 30 m^3/h and the heads (static and drawdown levels) are between 24 and 100 m. Therefore, the appropriate pump for these characteristics is the submersible pump. The diameter of the pipes for each pump has been calculated by an Excel sheet that used an iterative approach. Table 4.7 summarizes the diameters of the pipes and TDH for each well.

The total power of the eight submersible pumps equals 83.5 kW. Table 4.8 shows the selected pumps of the wells and their characteristics.

Each well has its own PV module that powers the inverter located close to the submersible pump. The capacity of PV power for all wells is 142.8 kW. Fig. 4.6 displays the single line diagram for Agadir well 1 with its specifications.

The configuration of the modules for each well is shown in Table 4.9.

The total capacity of all the inverters is 153 kW, which is matched with the PV modules in order to track the maximum power point. Table 4.10 shows the inverter models and their powers.

Table 4.7 Pipe diameters and TDH for each well.

TDH (m)	Friction loss in submain lines (m)	Diameter of submain lines (inch)	Friction loss in transmission line (m)	Diameter of transmission line (inch)	Well's name
61.13	8	2	8.41	5	Agadir well 1
83.32	34.95	1.5	17.24	4	Agadir well 2
79	30	1.5	21.47	5	Agadir well 3
60.78	15.36	2	10.68	6	Batha' well 1
85.1	31.2	2	11.5	6	Batha' well 2
56.48	10.64	2	10	6	Batha' well 3
154	40.8	1.5	29.1	5	Lemeileh well 1
119.21	13.1	1.5	8.99	5	Lemeileh well 2

Table 4.8 Eight pump models for each well.

Diameter of the discharge line (inch)	Motor's power (kW)	Pump's model	Well's name
2.5	5.5	SS19-09	Agadir well 1
2.5	7.5	SS19-12	Agadir well 2
2.5	7.5	SS19-12	Agadir well 3
3	11	SS30-09	Batha' well 1
3	11	SS30-09	Batha' well 2
3	7.5	SS30-06	Batha' well 3
3	18.5	SS27-19	Lmeileh well 1
3	15	SS27-15	Lmeileh well 2

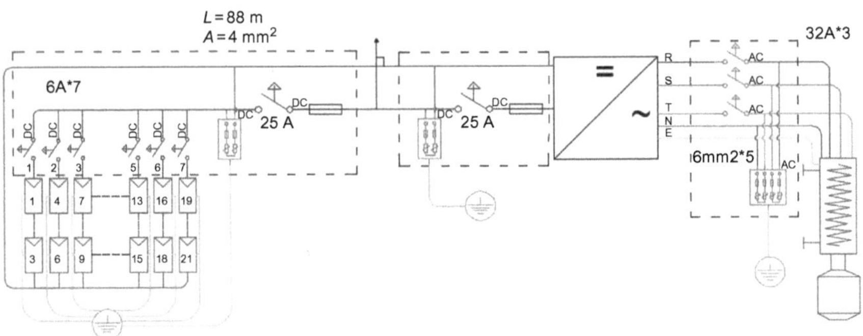

Fig. 4.6 Single line diagram of the Agadir well 1.

Table 4.9 Module configuration for each well.

Number of strings	Number of modules/strings	Total number of modules	Well's name
7	3	21	Agadir well 1
10	3	30	Agadir well 2
10	3	30	Agadir well 3
15	3	45	Batha' well 1
15	3	45	Batha' well 2
10	3	30	Batha' well 3
25	3	75	Lmeileh well 1
20	3	60	Lmeileh well 2

In this design example, the life cycle cost (LCC) is used as a method to evaluate the system economically. This approach gives a reflection of all the costs during the lifetime of the system. The initial costs (PV modules, inverter, pump, pipes, tanks, installation, etc.), annual maintenance, and

Table 4.10 Inverters powers and models.

Power capacity (kW)	Inverter's model number	Well's name
11	99,044,363	Agadir well 1
15	99,044,364	Agadir well 2
15	99,044,364	Agadir well 3
18.5	99,044,365	Batha' well 1
18.5	99,044,365	Batha' well 2
15	99,044,364	Batha' well 3
30	99,044,367	Lmeileh well 1
30	99,044,367	Lmeileh well 2

future replacement costs are analyzed for each well. The cost of the pumped water is priced in USD/m^3. There are many assumptions that have been taken into consideration in evaluating the LCC of the system. The economical parameters of discount rate, real discount rate, and inflation rate in Mauritania are assumed to be 13.82%, 8.07%, and 5.32% per year, respectively. Moreover, the maintenance cost and salvage cost are assumed to be 2% and 20%, respectively, of the initial cost per year. Replacement of the inverters and pumps is considered after 10 years of use.

Fig. 4.7 shows the costs of the system during its lifetime for each well.

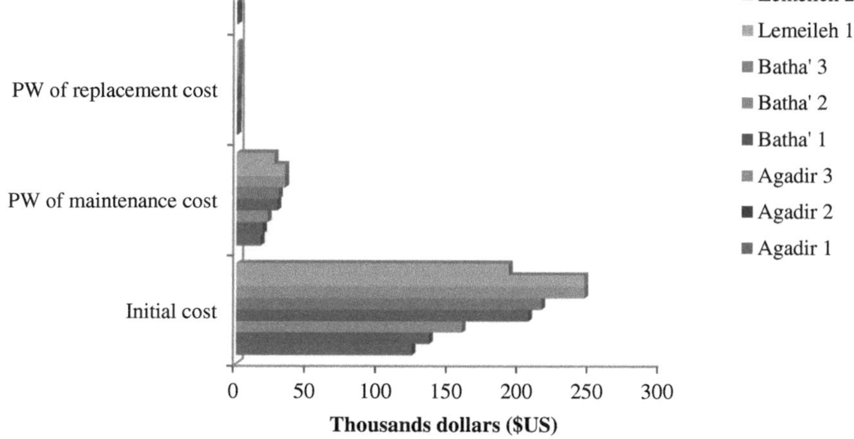

Fig. 4.7 Life cycle cost of PV pumping for each well.

According to the pumped water cost, the average cost of the eight wells equals 20.71 cents for each 1 m^3 of pumped water, which is low compared to the pumped water cost using diesel generators. The costs of the pumped water of the eight wells are between 14 and 43 cents per cubic meter. Lmeileh well 2 has the largest cost among the other wells. Conversely, Agadir well 1 has the lowest cost.

4.4 Numerical method for sizing photovoltaic water pumping systems

In general, the performance of any photovoltaic water pumping system depends on many factors, such as meteorological variables (i.e., solar radiation, ambient temperature, humidity), PV module specifications (i.e., conversion efficiency and tilt angle), and motor-pump–hydraulic system characteristics (i.e., I-V characteristic of the motor-pump set, and static, dynamic, and friction loss heads).

In general, three main methods can be used to size PV systems: the intuitive, numerical, and analytical methods. The simplest of the three is the intuitive method, which uses the worst month or the average monthly solar radiation in sizing the PV array and storage units. However, the intuitive method may lead to the over- or undersizing of the PV system, which consequently increases the cost or decreases the reliability of the system. In the analytical method, equations for the PV system size in terms of the reliability of the system can be developed and utilized. Using this method, the calculation of the system size becomes very simple and accurate, but the complexity of deriving the coefficient of the equations is the main disadvantage of the method.

By contrast, hourly meteorological data are used in the numerical method to simulate the system and describe its hourly energy flow. Afterward, the possible configurations of the system that achieve a specific level of reliability are found based on its performance. These configurations are then evaluated by applying an economic constraint to them to select the lowest cost system configuration. In numerical methods, the designers generally utilize the specific models of a system to simulate its performance based on meterological data. Afterward, the optimal sizing of the system is conducted iteratively based on an objective function that minimizes the cost of the system cost subject to a specific avaliabilty (i.e., loss of load probability or loss of power supply probability).

Due to the effective role of system configuration in the performance of a photovoltaic water pumping system, the optimal configuration should be used in the design of the system to meet the water demand. The numerical iterative method is used to obtain an optimal configuration for both the PV array and the storage tank. By this method, the configuration of the PV array (i.e., the number of modules connected in series and parallel) is specified directly in the proposed method based on the hourly weather condition for 1 year. By this method, the optimal configuration is specified based on the two objectives, which are the technical (reliability) and economic (cost) objectives, as illustrated in the following sections. The optimal configuration leads to a photovoltaic water pumping system with high reliability (i.e., minimum shortage time) and low cost. Fig. 4.8 shows the flowchart of the proposed sizing method.

The technical objective with such a method is represented by archiving the desired loss of load probability (LLP). The LLP can be defined as the ratio of the volume of deficit water to the volume of water demand within a period of time. By computing the deficit and water demand based on hourly solar radiation and ambient temperature, the LLP can then be estimated. The goal is to obtain a configuration that gives the minimum *LLP* value, which is given by,

$$LLP = \frac{\sum_{t=1}^{12*365} Q_d(t)}{\sum_{t=1}^{12*365} D(t)}. \tag{4.12}$$

On the other hand, in the numerical method, the price is also considered as an economical objective. The goal here is to find the configuration with the lowest cost from the set of configurations that satisfies a zero load rejection. A zero load rejection means that the *LLP* of the system is less than 0.01. The life cycle cost (*LCC*) is the most widely used to evaluate the cost of a photovoltaic water pumping system and consequently to choose the optimal configuration. The *LCC* of a photovoltaic water pumping system consists of the initial capital cost, the present value of the maintenance cost, and the present value of the replacement cost:

$$LCC = IC + MC + RC, \tag{4.13}$$

where *IC* is the initial capital cost (USD), *MC* is the present value of the maintenance cost (USD), and *RC* is the present value of the replacement cost (USD). These terms are discussed in detail in the following sections.

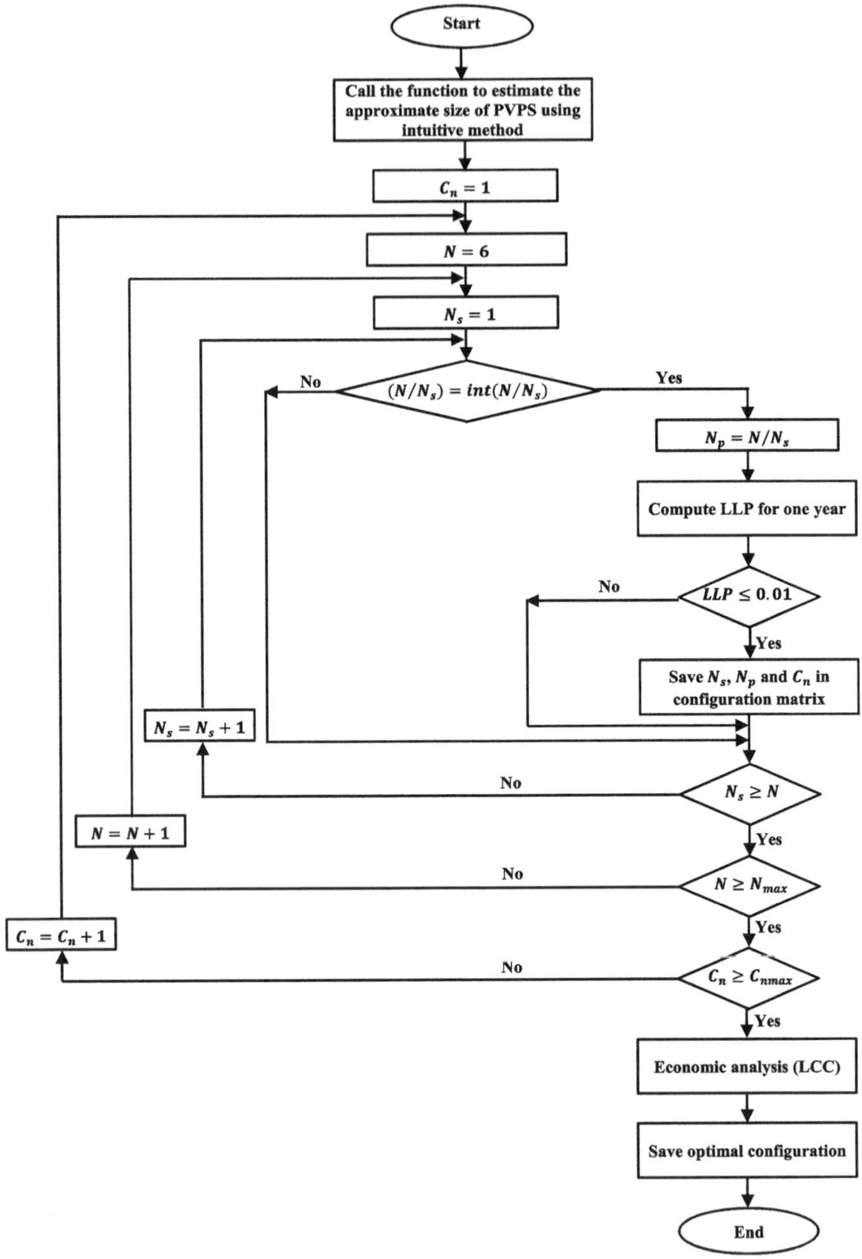

Fig. 4.8 Flowchart of the numerical method for sizing a photovoltaic water pumping system.

The initial capital cost of a PVPS consists of the price of the system's components, the cost of civil works, and the expenditure for the system's design and installation. The initial capital cost of the PVPS can be expressed by

$$IC = CA_{PV} + UC_{PV} + CA_C + UC_C + CA_{MP} + UC_{MP} + CA_T + UC_T + ICI, \tag{4.14}$$

where CA_i is the capacity of the ith component of PVPS, UC_i is the cost per unit of the ith component (USD/unit), and ICI is the total constant cost, including the cost of installation and civil works (USD).

As for the maintenance *value,* the present value of the maintenance cost of the photovoltaic water pumping system is given by

$$MC_r = \begin{cases} MC_{0r}{*}\left(\dfrac{1+FR}{IR-FR}\right)\left(1-\left(\dfrac{1+FR}{1+IR}\right)^{LP}\right) & \text{if } IR \neq FR \\ MC_{0r}{*}LP & \text{if } IR = FR \end{cases} \tag{4.15}$$

The maintenance cost of the system's component in the first year can be expressed as a percentage of the initial capital cost of the components:

$$MC_{0r} = k_r{*}IC_r, \tag{4.16}$$

The total maintenance cost of the photovoltaic water pumping system is given by

$$MC = \sum_{r}^{3} MC_r, \tag{4.17}$$

where FR is the annual inflation rate, IR is the annual interest rate, LP is the lifetime of the photovoltaic water pumping system (year), MC_{0r} is the maintenance cost of the rth component in the first year (USD), MC_r is the maintenance cost of the rth component (USD), IC_r is the initial cost of the rth component (USD), k_r is a constant that refers to the maintenance cost as a percentage of the initial capital cost of the rth component, and the r values are 1, 2, and 3, which are equivalent to the PV array, the motor-pump set, and the storage tank, respectively.

Finally, in a photovoltaic water pumping system, the PV array is a component with a longer lifetime. Thus, the lifetime of the PVPS is often considered equal to the lifetime of the PV array. By contrast, the lifetime of the DC-DC converter and that of the motor-pump set are less than that of the system. Thus, the converter and motor-pump set require periodic replacement over the

Table 4.11 Financial data of a system's components.

Seq.	Component	Cost/unit	Life time (year)	k_r	N_r	LP	IR	FR
1	PV	1 USD/Wp	20	1%	0	20	8%	4%
2	Converter	0.5 USD/W	10	0%	1			
3	Motor-pump set	0.75 USD/W	10	3%	1			
4	Storage tank	20 USD/m^3	20	1%	0			

lifetime of the system. The present value of the replacement cost of the photovoltaic water pumping system is given by

$$RC_k = IC_k * \sum_{j=1}^{N_r} \left(\frac{1+FR}{1+IR}\right)^{\left(\frac{LP*j}{N_r+1}\right)}, \tag{4.18}$$

$$RC = \sum_{k=1}^{2} RC_k, \tag{4.19}$$

where IC_k is the initial cost of the kth component (USD), RC_k is the replacement cost of the kth component (USD), N_r is the number of component replacements over the lifetime of the system, and the k values are 1 and 2, which are equivalent to the converter and motor-pump set, respectively. Table 4.11 shows the assumed financial data of a system's components.

4.4.1 Design example for a photovoltaic water pumping system using the numerical method for a location in Malaysia

In this example, the proposed methodology is applied to size and analyze the performance of a photovoltaic water pumping system using the optimal configuration to supply water for irrigation and drinking purposes in a village of 120 people in Kuala Lumpur, Malaysia. The load profile is considered to be constant with a total daily water demand of 30 m^3.

The hourly data for 1 year of ambient temperature and solar radiation on the horizontal plane are used in the sizing and performance test of the photovoltaic water pumping system. An iterative numerical method is used to choose the optimal configuration based on the technical and economic objectives. The range of the number of PV modules used in the numerical method is chosen intuitively from six to 32 modules. The range of the size of the storage tank is chosen from 1 to 120 m^3. For a specific number of PV modules, all the possible PV array configurations are tested based on the hourly data for 1 year to compute the *LLP* for the proposed configuration. Thus, both the number of modules connected in series and those in parallel are specified directly based on a specific reliability and minimum cost conditions. The following code is applied using MATLAB software based on the flowchart illustrated in Fig. 4.8.

```
%%%%%%%%% Sizing PVPS for one year based on numerical iterative
%%%%%%%%% method. To compute LLP only Size of tank is changed from 1m^3
to 160m^3 with
%%%%%%%%% 1m^3 as a step size. In the meanwhile, the PV array size is
%%%%%%%%% changed from 6 to 32 modules. For 20m as a head
%%%%%%%%% The computations are hourly for one year data
%%%%%%%%% Created in: 28/8/2015
%%%%%%%%% Modified in: 23/11/2015
clc; clear all; close all;
t=cputime; %To specify the consumed time to compute LLP for all
configurations
t1=cputime; %To specify the consumed time for sizing program
CNs=zeros(1,80000);
%Create a matrix for Ns values for each configuration realizes
LLP<=0.01 over a year
CNp=zeros(1,80000);
%Create a matrix for Np values for each configuration realizes
LLP<=0.01 over a year
CCn=zeros(1,80000);
%Create a matrix for Cn values for each configuration realizes
LLP<=0.01 over a year
CLLP=ones(1,80000);
%Create a matrix for LLP values for each configuration realizes
LLP<=0.01 over a year
CLLP=CLLP.*(-0.5);
CQexcess=ones(1,80000);
%Create a matrix for Qexcess values for each configuration realizes
LLP<=0.01 over a year
CQexcess=CQexcess.*(-0.5);
CQdeficit=ones(1,80000);
```

Continued

—cont'd

```
%Create a matrix for Qexcess values for each configuration realizes
LLP<=0.01 over a year
CQdeficit=CQdeficit.*(-0.5);
CQ=ones(1,80000);
%Create a matrix for Q values for each configuration realizes LLP<=0.01
over a year
CQ=CQ.*(-0.5);
q=0;
%Initialize the counter for indexing CNs, CNp, CCn & LLP matrices
G=xlsread('Result_Sizing_Numerical_Iterative_Method.xlsx',1,'I10:
I4389');
%Reading the hourly solar radiation (W)
Tc=xlsread('Result_Sizing_Numerical_Iterative_Method.xlsx',1,'J10:
J4389');
%Reading the hourly cell temperature (K)
Vm=xlsread('Result_Sizing_Numerical_Iterative_Method.xlsx',1,'L10:
L4389');
%Reading the hourly voltage of one module (V)
Im=xlsread('Result_Sizing_Numerical_Iterative_Method.xlsx',1,'M10:
M4389');
%Reading the hourly current of one module (A)
for Cn=1:120                %Size of storage tank is increased by 1m^3
    for N=6:32              %Number of modules are increased by one
        for Ns=1:N          %Number of series modules are increased by one
            if rem(N/Ns,1)==0
                Np=N/Ns;        %To set the number of parallel modules
                %%%%%% COMPUTE LLP %%%%%%%%%%
                %% h1 and h2 are changed according to pumping head %%
                %%h1=20, 30, 40, 50 and 60, respectively.
                %%h2=0.1444,  0.1907,  0.2371,  0.2835  and  0.3298,
                respectively
                h1=20;  %%Factors of head equation we have got it from
                head calculations
                h2=0.1444; %%%&&&Factors of head equation we have got
                it from head calculations
    %%%%%%%%%%%%%%%%%%%%%%%%%%%%%% PV ARRAY %%%%%%%%%%%%%%%%%%%%%%%%%%
    %%%%%%%%
                Va=Ns.*Vm;      %Computing the hourly voltage of PV array (V)
                Ia=Np.*Im;      %Computing the hourly current of PV array (A)
                Vpv=Va;         %Output voltage of PV array for storing
                                purpose
```

Continued

—cont'd

```
            Ipv=Ia;           %Output current of PV array for
                               storing purpose
            Pao=Va.*Ia;        %Computing the hourly output power
                               of PV array (W)
            Am=0.9291;         %area of PV module (m^2)
            A=Ns*Np*Am;         %Computing the area of PV array (m^2)
            Pai=A.*G;            %Computing the hourly input power
                                 of PV array (W)
            effa=Pao./Pai;        %Computing the hourly efficiency
                                  of PV array
    %%%%%%%%%%%%%%%%%%%%%%%%%%%%%%% MOTOR %%%%%%%%%%%%%%%%%%%%%%%%%%%%%%%
%%%%%%
    effconv=0.9;    %DC-DC converter efficiency
    Va=0.95.*Va;     %The output voltage of DC-DC converter
    Ia=0.90.*Ia;      %The output current of DC-DC converter
    Ra=0.8;            %Armature resistance of DC motor (Ohm)
    Km=0.175;           %Torque and back emf constant (V/(rad/sec))
    Ebb=Va-(Ra.*Ia); %Computing the hourly back emf voltage of motor
(V)
    %%%%%%%%%% The case of overcurrent supplied to motor by PV array %%
%%%%%%
    Eb=(Ebb>=0).*Ebb; %Set Eb=0 when Ebb<0 (overcurrent)/turn off motor
    Ia=(Ebb>=0).*Ia;   %Set Ia=0 when Ebb<0 (overcurrent)/turn off motor
    Va=(Ebb>=0).*Va;    %Set Va=0 when Ebb<0 (overcurrent)/turn off motor
    %%%%%%%%%%%%%%%%%%%%%%%%%%%%%%%%%%%%%%%%%%%%%%%%%%%%%%%%%%%%%%%%%%%%%%
%%%%%%%%
    Tm=Km.*Ia;    %Computing the hourly torque of DC motor
    Tmm=(Tm==0).*1;
    Tm1=Tmm+Tm;
    Rou=1000;      %Density of water (Kg/m^3)
    g=9.81;          %Acceleration due to gravity (m/Sec^2)
    d1=33.5*0.001;   %Inlet impeller diameter (mm)
    d2=160*0.001;     %outlet impeller diameter (mm)
    beta1=38*2*pi/360;%Inclination angle of impeller blade at impeller
inlet (degree)
    beta2=33*2*pi/360;%Inclination angle of impeller blade at impeller
outlet(degree)
    b1=5.4*0.001;                   %Height of impeller blade at
                                     impeller inlet (mm)
    b2=2.2*0.001;                   %Height of impeller blade at
                                     impeller outlet (mm)
```

Continued

—cont'd

```
    Kp=Rou*2*pi*b1*(d1/2)^2*tan(beta1)*((d2/2)^2   ((b1*(d1/2)^2*tan
(beta1))/(b2*tan(beta2))));%Computing the hourly output power of DC motor (W)
    Pdev=Eb.*Ia;
    Omega=abs(sqrt((Km.*Ia)./Kp));
    %Computing the hourly angular speed of motor (rad/sec)
    Pmo=Pdev;
    Pmi=Pao.*0.9; %Computing the hourly input power of DC motor (W)
    PMI1=(Pmi==0).*1; %To overcome divided by zero
    PMI2=Pmi+PMI1;  %To overcome divided by zero
    effm=Pmo./PMI2;  %Computing the hourly efficiency of DC motor
    %%%%%%%%%%%%%%%%%%%%%%%%%%%%%% PUMP %%%%%%%%%%%%%%%%%%%%%%%%%%%%%%
%%%%%
    Tp=Tm; %The produced torque by motor is equal the torque required
for pump (Nm)
    Eh=Tp.*Omega; %Computing the hydraulic energy (W)
    Ppo=Eh;       %Computing the hourly output power of pump (W)
    Ppo=(Ebb>=0).*Ppo;
    Ppi=Pmo;                %Computing the hourly input power of pump (W)
    PPI1=(Ppi==0).*1;       %To overcome divided by zero
    PPI2=Ppi+PPI1;          %To overcome divided by zero
    effpp=Ppo./PPI2;        %Computing the hourly efficiency of pump
    effp=(effpp<=0.95).*effpp;
    Q=zeros(length(Eh),1);
    for ii=1:length(Eh)
        r1=h2*2.725;          %Computing the flow rate of water
        r2=0;                 %Computing the flow rate of water
        r3=h1*2.725;          %Computing the flow rate of water
        r4=-Eh(ii);           %Computing the flow rate of water
        r=roots([r1 r2 r3 r4]);        %Computing the flow rate of water
        if (imag(r(1))==0 && real(r(1))>0)
     %Choosing the real value of the flow rate of water
            QQQ=real(r(1));
      elseif (imag(r(2))==0 && real(r(2))>0)
     %Choosing the real value of the flow rate of water
            QQQ=real(r(2));
      elseif (imag(r(3))==0 && real(r(3))>0)
     %Choosing the real value of the flow rate of water
            QQQ=real(r(3));
      else
            QQQ=0;
```

Continued

—cont'd

```
    %If all the roots are complex and/or the real part is negative
number or zero
    end
    QQ(ii,1)=QQQ; %Hourly flow rate (m^3/h)
   end
   Q1=(QQ==0).*1;   %To overcome divided by zero
   Q2=QQ+Q1;        %To overcome divided by zero
   H=Eh./(2.725.*Q2); %Computing the head of pumping water (m)
   %%%%%%%%%%% OVERALL SYSTEM %%%%%%%%%%%%%%%%%%%%%%%%%%%%%%%%%%%%%
%%%%%%%%
   effsub=effm.*effp.*   effconv; %Computing   hourly   subsystem
                                  efficiency
   effoverall=effa.*effm.*effp.* effconv; %Computing hourly overall
                                          efficiency
   QQ=(Ebb>=0).*QQ;
   QQ=(effpp<=0.95).*QQ;
   Q=[0;QQ];        %To add initial case Q=0 for programming purposes
   d=2.5;
   %Hourly demand water (m^3/h)
   Cr=zeros(length(Q),1);
   %To specify the size of current resident matrix of storage tank
   Qexcess_pv=zeros(length(Q),1); %To  specify  the  size  of  excess
water matrix
   Qexcess_s=zeros(length(Q),1);
   SOC=zeros(length(Q),1);
   Qdef_pv=zeros(length(Q),1);
   %To specify the size of deficit water matrix (before tank)
   Qdeficit_s=zeros(length(Q),1);
   %To specify the size of deficit water matrix (after tank)
   X=Q(2:end,1)-d;      %Difference  between  the  hourly  production
and demand water
   %%%%%%%%%%%%%%********** Before Tank **********%%%%%%%%%%%%%%%
   Qdef_pv(2:end,1)=(X<0).*abs(X); % hourly deficit water before tank
(m^3)
   %%%%%%%%%%%%%%********** After Tank **********%%%%%%%%%%%%%%%
   C=length(Q)-1;
   Qexcess_pv(2:end,1)=(X>=0).*abs(X); %hourly  excess  water  before/
after tank (m^3)
   for i=1:C
       Cr(i+1,1)=((Cr(i,1)+X(i,1))>=0).*abs(Cr(i,1)+X(i,1));
     %To compute the hourly current resident water in the tank (m^3)
```

Continued

—cont'd

```
    SOC(i+1,1)=Cr(i+1,1)/Cn;
    %Computing the hourly state of charge of storage tank
    if SOC(i+1,1)>=1
        SOC(i+1,1)=1;
        Qexcess_s(i+1,1)=Cr(i+1,1)-Cn;
        Cr(i+1,1)=Cn;
    else
        Qexcess_s(i+1,1)=0;
    end
    Qdeficit_s(i+1,1)=((Cr(i,1)+X(i,1))<0).*abs(Cr(i,1)+X(i,1));
    %To compute the hourly deficit water (m^3/h) (after tank)
end
Q=Q(2:end,1);   %Final computing of hourly flow rate of water (m^3)
sumQ=sum(Q);    %To sum the hourly Q values over a year
Qexcess_pv=Qexcess_pv(2:end,1);
%Final computing of hourly excess water before and after tank (m^3)
Qexcess_s=Qexcess_s(2:end,1);
%Final computing of hourly excess water after the tank is filled (m^3)
Qexcess=sum(Qexcess_s);
Qdeficit_s=Qdeficit_s(2:end,1);
%Final computing of hourly deficit water after tank (m^3)
Qdeficit=sum(Qdeficit_s);
Qdef_pv=Qdef_pv(2:end,1);
%Final computing of hourly deficit water before tank (m^3)
Cres=Cr(2:end,1);
%Final computing of hourly current resident water in tank (m^3)
SOC=SOC(2:end,1);
%Final computing of hourly state of charge (SOC)
D=zeros(length(Q),1)+d; %Constructing the matrix of hourly demand
 water (m^3)
LLPh=Qdeficit_s(1:end,1)./D(1:end,1);    %Computing the hourly LLP
LLP=sum(Qdeficit_s(1:end,1))/sum(D(1:end,1)); %Computing the LLP
                                                of one year
if LLP<=0.01
    q=q+1;   %Increment the index of CNs, CNp, CCn & LLP matrices
    CNs(1,q)=Ns;  %Store the value of Ns
    CNp(1,q)=Np; %Store the value of Np
    CCn(1,q)=Cn;  %Store the value of Cn
    CLLP(1,q)=LLP; %Store the value of LLP
    CQexcess(1,q)=Qexcess;  %Store the value of excess water
    CQdeficit(1,q)=Qdeficit;  %Store the value of deficit water
    CQ(1,q)=sumQ;     %To store the value of Q over a year
end
```

Continued

—cont'd

```
    end
   end
  end
end
C=[CNs; CNp; CCn];
%Matrix with all configurations
%%%%%% NEGLECTING THE SURPLUS COLUMNS IN MATRIX (C), THE COLUMNS WITH
ZERO VALUES %%%%%
idx=C(1,:)==0; %Index those columns which have a zero value in the
first row
C=C(:,~idx); %Take all rows, but only columns that do not have a zero
value in the first columns
iidx=CLLP(1,:)==-0.5; %Index those columns which have a -0.5 value
CLLP=CLLP(1,~iidx); %Take all values, but only columns that do not
have a -0.5 value
CQexcess=CQexcess(1,~iidx);
CQdeficit=CQdeficit(1,~iidx);
CQ=CQ(1,~iidx);
e=cputime-t %To compute the total time consumed for computing LLP
program for all configurations
%%%%%% COST COMPUTATION FOR ALL CONFIGURATIONS THOSE SATISFY LLP<=0.01%
%%%%%%%
CAc=800;               %Total capacity of converter required for system (W)
CAmp=840;               %Total capacity of motor-pump set required for
                         system (W)
UCpv=1;                  %%Unit cost of PV ($/Wp)
UCc=0.5;                  %%Unit cost of converter ($/W)
UCmp=0.75;                 %%Unit cost of motor-pump set ($/W)
UCt=20;                     %%Unit cost of storage tank ($/m^3)
ICI=4000;                    %%Civil and installation works cost ($)
FR=0.04;                      %%Inflation rate
IR=0.08;                       %%Interest rate
LP=20;                          %%Life time of PVPS
Nr=1;                            %%Number of replacement times for
motor-pump set and converter
u=length(C);                %Specifying the number of configurations
those satisfy LLP<=0.01
LCC_cost=zeros(1,u);
%Creating matrix to store the cost of all configurations those satisfy
LLP<=0.01
```

Continued

—cont'd

```
cost_Year=zeros(1,u);   %Creating matrix to store the yearly cost of all configurations
COU_vector=zeros(1,u);  %Creating matrix to store the water unit cost of all configurations
for p=1:u
    Ns=C(1,p);             %Taking each configuration to compute its cost
    Np=C(2,p);             %Taking each configuration to compute its cost
    Cn=C(3,p);             %Taking each configuration to compute its cost
    N=Ns*Np;               %Total number of PV modules
    %%%%%%%% COST SCRIPT %%%%%%%%%%%%%%%%%%%%
    CApv=120*N;            %Total capacity of PV required for system (Wp)
    CAt=Cn;                %Total capacity of storage tank
                            required for system (m^3)
 %%%%%%%%%%%%%%%%%%% COMPUTING INITIAL COST OF PVPS %%%%%%%%%%%%%%%%%%%%%
 IC=(CApv*UCpv)+(CAc*UCc)+(CAmp*UCmp)+(CAt*UCt)+ICI;   %Initial cost of PVPS ($)
 %%%%%%%%%%%%%%%%%%% COMPUTING REPLACEMENT COST OF PVPS %%%%%%%%%%%%%%%%%
 RCCmp=zeros(1,2);
 %Array to accumalate the replacement cost for every replacement of motor-pump set
 RCCc=zeros(1,2);
 %Array to accumalate the replacement cost for every replacement of converter
 for j=1:Nr
     RRCmp=(CAmp*UCmp)*(((1+FR)/(1+IR))^((LP*j)/(Nr+1)));
     %Replacement cost of motor-pump set ($)
     RRCc=(CAc*UCc)*(((1+FR)/(1+IR))^((LP*j)/(Nr+1)));
     %Replacement cost of motor-pump set ($)
     RCCmp(1,j)=RRCmp;
     %Accumalate the replacement costs for motor-pump set
     RCCc(1,j)=RRCc;
     %Accumalate the replacement costs for converter
 end
 RCmp=sum(RCCmp);
 %Sum the replacement cost for L replacement motor-pump set ($)
 RCc=sum(RCCc);
 %Sum the replacement cost for L replacement converter ($)
 RC=RCmp+RCc;
 %Computing the replacement cost for PVPS ($)
```

Continued

—cont'd

```
%%%%%%%%%%%%%%%%% COMPUTING OPERATION AND MAINTENANCE COST OF PVPS %%%
%%%%%
 MCpv0=0.01*(CApv*UCpv); %Maintenance and operation cost of PV in the
first year ($)
 MCmp0=0.03*(CAmp*UCmp);
 %Maintenance and operation cost of motor-pump set in the first year ($)
 MCt0=0.01*(CAt*UCt);
 %Maintenance and operation cost of storage tank in the first year ($)
 MCpv=MCpv0*((1+FR)/(IR-FR))*(1-((1+FR)/(1+IR))^LP);
 %Maintenance and operation cost of PV along life time of PVPS ($)
 MCmp=MCmp0*((1+FR)/(IR-FR))*(1-((1+FR)/(1+IR))^LP);
 %Maintenance and operation cost of motor-pump set along life time of PVPS ($)
 MCt=MCt0*((1+FR)/(IR-FR))*(1-((1+FR)/(1+IR))^LP);
 %Maintenance and operation cost of storage tank along life time of PVPS ($)
 MC=MCpv+MCmp+MCt;
 %Calculating the maintenance and operation costs of PVPS along life
time of PVPS ($)
 %%%%%%%%%%%%%%%%% COMPUTING LIFE CYCL COST OF PVPS %%%%%%%%%%%%%%%%%
%%%
 LCC=IC+RC+MC;
 %Computing the LCC of PVPS ($)
 %%%%%%%%%%%%%%%%% COMPUTING COST OF ENERGY %%%%%%%%%%%%%%%%%%%%%%%%%
%%%%
 LCC_Year=LCC/LP;
 %Computing the system cost for one year ($/year)
 %Water_Volume_Year=10585
 %Computing the volume of pumped water per year (m^3/year)
 %COE=LCC_Year/Water_Volume_Year;
 %Computing the cost of energy ($/m^3)
 LCC_cost(1,p)=LCC;
 %To store the LCC for each configuration
 cost_Year(1,p)=LCC_Year;
 %To store the yearly system cost for each configuration
 COU=LCC_Year/(CQ(1,p)-CQexcess(1,p));
 %Cost of water unit ($/m^3)
 COU_vector(1,p)=COU;
end
% %%%%%%% SPECIFYING THE CONFIGURATION THAT SATISFY LLP<=0.01 WITH
MINIMUM COST %%%%%%%
[n,i]=min(LCC_cost); %To specify the minimum cost configuration (value
and index)
cost_optimal=n;  %LCC for optimal configuration
```

Continued

—cont'd

```
cost_Year_optimal=n/LP; %yearly system cost for optimal configuration
COU_optimal=COU_vector(1,i);         %Water unit cost value of optimal
                                      configuration
Ns_optimal=C(1,i);                    %Optimal Ns value
Np_optimal=C(2,i);                     %Optimal Np value
N_optimal =Ns_optimal*Np_optimal;     %To compute the total number of
                                       PV modules for optimal size
Cn_optimal=C(3,i);                      %Optimal Cn value
LLP_optimal=CLLP(1,i);                   %Optimal LLP value
Qexcess_optimal=CQexcess(1,i);            %Qexcess for one year of
                                           optimal PVPS configuration
Qdeficit_optimal=CQdeficit(1,i);           %Qdeficit for one year of
                                            optimal PVPS configuration
Q_optimal=CQ(1,i);                          %Q for one year of optimal
                                             PVPS configuration
%%%%%%%%% SAVING RESULTS IN EXCEL SHEET %%%%%%%%%%%%%%%%%%%%%%%%%
%%%%
%%%%%%%%% Optimal Configuration %%%%%%%%%%%%%%%%%%%%%%%%%%%%%%%%
%%%%%
xlswrite('Result_Sizing_Numerical_Iterative_Method.xlsx',
N_optimal,2,'N19');
%To save N_optimal in excel sheet
xlswrite('Result_Sizing_Numerical_Iterative_Method.xlsx',
Ns_optimal,2,'O19');
%To save Ns_optimal in excel sheet
xlswrite('Result_Sizing_Numerical_Iterative_Method.xlsx',
Np_optimal,2,'P19');
%To save Np_optimal in excel sheet
xlswrite('Result_Sizing_Numerical_Iterative_Method.xlsx',
Cn_optimal,2,'Q19');
%To save Cn_optimal in excel sheet
xlswrite('Result_Sizing_Numerical_Iterative_Method.xlsx',
LLP_optimal,2,'R19');
%To save M_optimal in excel sheet
xlswrite('Result_Sizing_Numerical_Iterative_Method.xlsx',
COU_optimal,2,'X19');
%To save M_optimal in excel sheet
xlswrite('Result_Sizing_Numerical_Iterative_Method.xlsx',
cost_optimal,2,'S19');
%To save cost_optimal (LCC) in excel sheet
xlswrite('Result_Sizing_Numerical_Iterative_Method.xlsx',
cost_Year_optimal,2,'T19');
%To save cost_optimal (LCC) in excel sheet
```

Continued

—cont'd

```
xlswrite('Result_Sizing_Numerical_Iterative_Method.xlsx',
Qexcess_optimal,2,'U19');
%To save Qexcess for optimal configuration in excel sheet
xlswrite('Result_Sizing_Numerical_Iterative_Method.xlsx',
Qdeficit_optimal,2,'V19');
%To save Qdeficit for optimal configuration in excel sheet
xlswrite('Result_Sizing_Numerical_Iterative_Method.xlsx',
Q_optimal,2,'W19');
%To save Q for optimal configuration in excel sheet
%%%%%%%%%% All Configurations%%%%%%%%%%%%%%%%%%%%%%
CN=C(1,:).*C(2,:); %To compute the matrix of the total number of PV modules
CN_transpose=CN'; %To convert from row vector to column vector to store
it in excel sheet
CNs_transpose=C(1,:)'; %To convert from row vector to column vector to
                         store it in excel sheet
CNp_transpose=C(2,:)';  %To convert from row vector to column vector
                          to store it in excel sheet
CCn_transpose=C(3,:)';   %To convert from row vector to column vector
                           to store it in excel sheet
CLLP_transpose=CLLP(1,:)'; %To convert from row vector to column vector
to store it in excelsheet
CQexcess_transpose=CQexcess';
CQdeficit_transpose=CQdeficit';
LCC_cost_transpose=LCC_cost';
%To convert from row vector to column vector to store it in excel sheet
cost_Year_transpose=cost_Year';
%To convert from row vector to column vector to store it in excel sheet
COU_vector_transpose=COU_vector';
%To convert from row vector to column vector to store it in excel sheet
xlswrite('Result_Sizing_Numerical_Iterative_Method.xlsx',
CN_transpose,2,'AA2');
%To save all N values those satisfy LLP<=0.01 in excel sheet
xlswrite('Result_Sizing_Numerical_Iterative_Method.xlsx',
CNs_transpose,2,'AB2');
%To save all Ns values those satisfy LLP<=0.01 in excel sheet
xlswrite('Result_Sizing_Numerical_Iterative_Method.xlsx',
CNp_transpose,2,'AC2');
%To save all Np values those satisfy LLP<=0.01 in excel sheet
xlswrite('Result_Sizing_Numerical_Iterative_Method.xlsx',
CCn_transpose,2,'AD2');
%To save all Cn values those satisfy LLP<=0.01 in excel sheet
xlswrite('Result_Sizing_Numerical_Iterative_Method.xlsx',
CLLP_transpose,2,'AE2');
```

Continued

—cont'd

```
%To save all LLP values those satisfy LLP<=0.01 in excel sheet
xlswrite('Result_Sizing_Numerical_Iterative_Method.xlsx',
LCC_cost_transpose,2,'AF2');
%To save all LCC for configurations those satisfy LLP<=0.01 in excel sheet
xlswrite('Result_Sizing_Numerical_Iterative_Method.xlsx',
CQexcess_transpose,2,'AH2');
%To save all Qexcess values those satisfy LLP<=0.01 in excel sheet
xlswrite('Result_Sizing_Numerical_Iterative_Method.xlsx',
CQdeficit_transpose,2,'AI2');
%To save all Qdeficit values those satisfy LLP<=0.01 in excel sheet
xlswrite('Result_Sizing_Numerical_Iterative_Method.xlsx',
cost_Year_transpose,2,'AG2');
%To save yearly system cost for configurations in excel sheet
xlswrite('Result_Sizing_Numerical_Iterative_Method.xlsx',
COU_vector_transpose,2,'AJ');
%To save yearly system cost for configurations in excel sheet
el=cputime-tl
%To compute the total time consumed for executing sizing program
xlswrite('Result_Sizing_Numerical_Iterative_Method.xlsx',e,2,'Y2');
%To save time consumed for computing LLP for all configurations in excel sheet
xlswrite('Result_Sizing_Numerical_Iterative_Method.xlsx',el,2,'Z2');
%To save the total time consumed for executing sizing program in excel sheet
```

Based on the executed code, the optimal PV array configuration is four modules connected in parallel and five modules connected in series. The correlation between the size of the PV array and the *LCC* for the configurations with an *LLP* less than or equal to 0.01 is shown in Fig. 4.9. The figure clearly shows that the PVPS consisting of 20 modules offers the lowest cost with an *LLP* ranging from 0.0035 to 0.01. Fig. 4.10 shows the sizes of the storage tank that lead to an *LLP* less than or equal to 0.01. According to the figure, the effective size of the storage tank is from 39 to 80 m^3. The *LLP* value is not affected when the size of the tank is greater than 80 m^3. A 50 m^3 storage tank is used in the PVPS to exhibit 0.005572 as the *LLP* of the system for 1 year and also to reduce the volume of excess water. The correlation between the number of PV modules and the LCC is depicted in Fig. 4.11, where the PV array with 20 modules offers the lowest *LCC*, which is about USD 9865.

The operation performance of a PVPS based on the optimal configuration in March, which has the highest average daily monthly solar radiation of about 5131.552 W/m^2, is presented in Fig. 4.12. The first part of the figure illustrates the daily solar radiation. The second part shows the energy

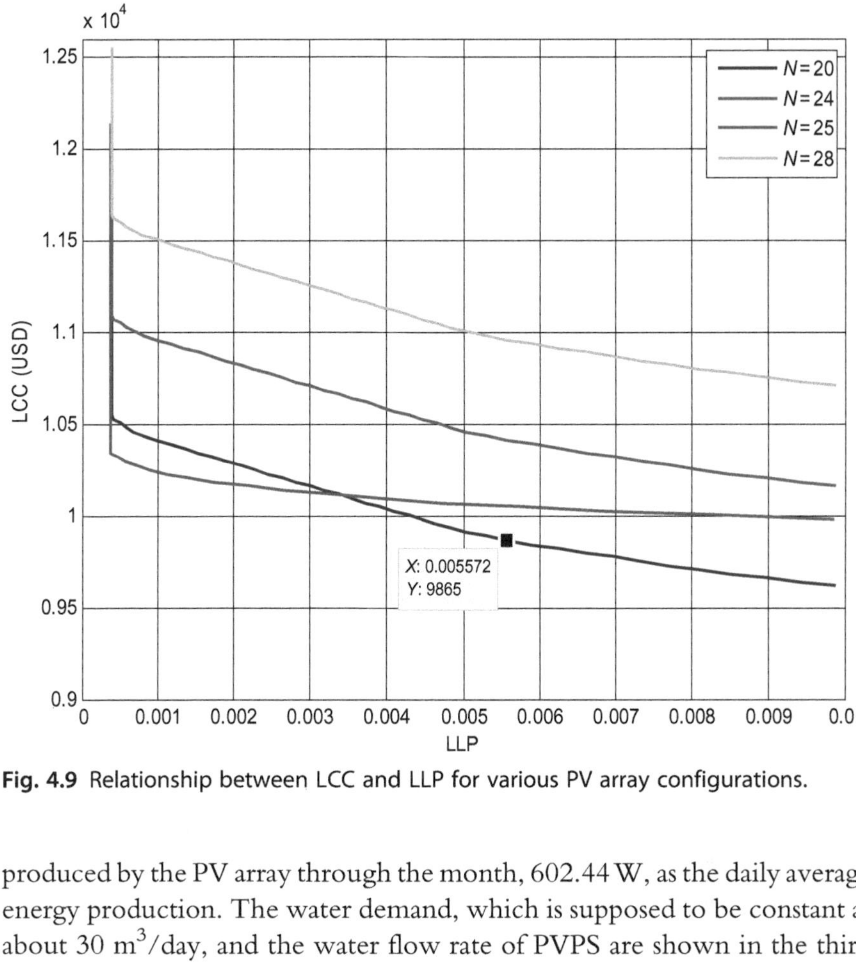

Fig. 4.9 Relationship between LCC and LLP for various PV array configurations.

produced by the PV array through the month, 602.44 W, as the daily average energy production. The water demand, which is supposed to be constant at about 30 m^3/day, and the water flow rate of PVPS are shown in the third part of the figure. The system productivity without the storage tank often meets the water demand per day in March, when the lowest average daily water pumped by the system is on March 15 at around 30.151 m^3. The fourth and sixth parts of the figure explain the high reliability of the PVPS design, where the daily *LLP* and Q_d are equal to zero in all the days of March. The fifth part of the figure explains the daily excess water throughout the month. The last part of the figure shows the daily state of charge of the storage tank. In most days of March, the *SOC* of the storage tank is more than 80% because the productivity of the system meets the water demand.

The performance of the PVPS in December, which has the lowest average daily monthly solar radiation of about 3779.865 W/m^2, is shown in Fig. 4.13. The daily average energy produced by the PV array through the month is 425.038 W. The system productivity without the storage tank fluctuates about

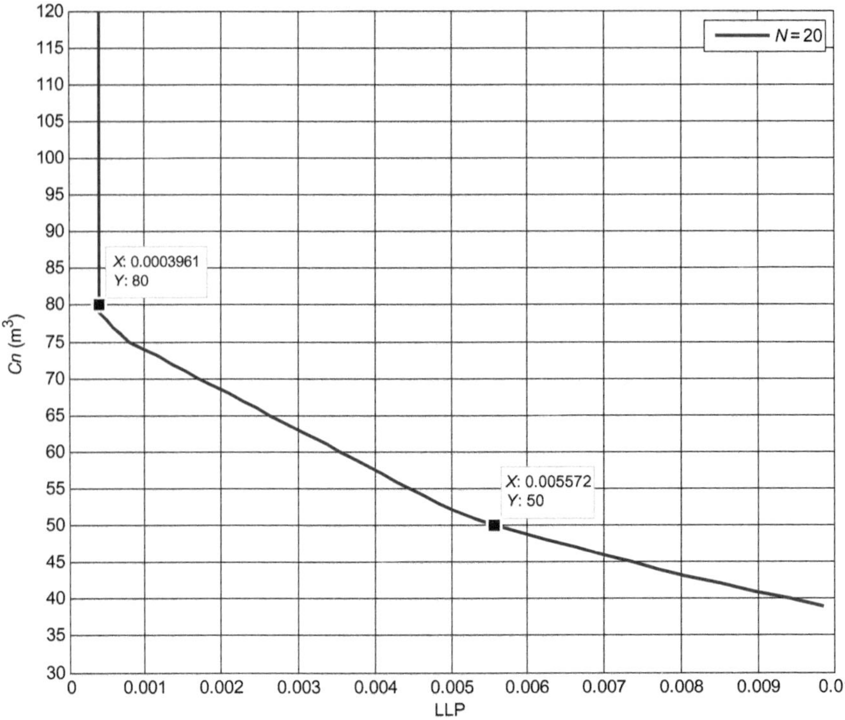

Fig. 4.10 Relationship between LLP and the size of a storage tank for a 20-module PV array configuration.

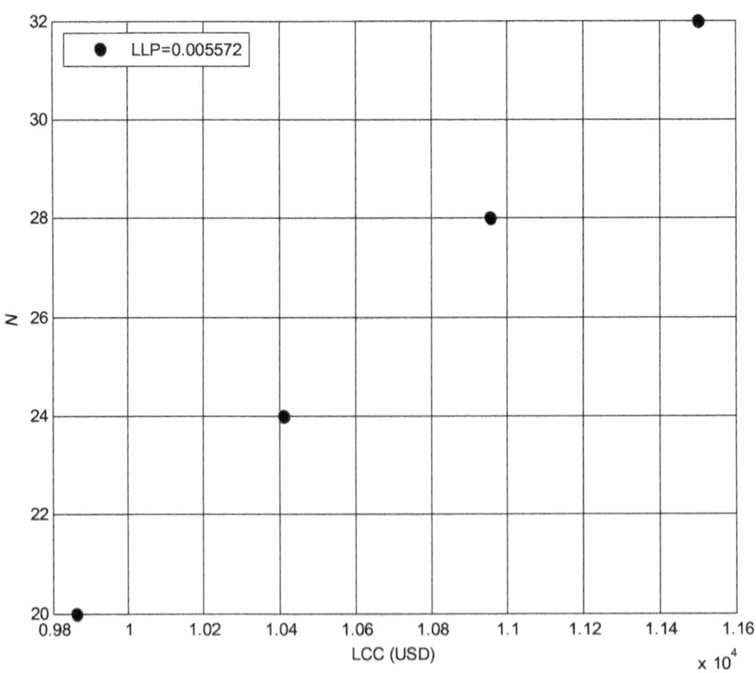

Fig. 4.11 Relationship between the LCC and PV array configurations that satisfied a -specific LLP value.

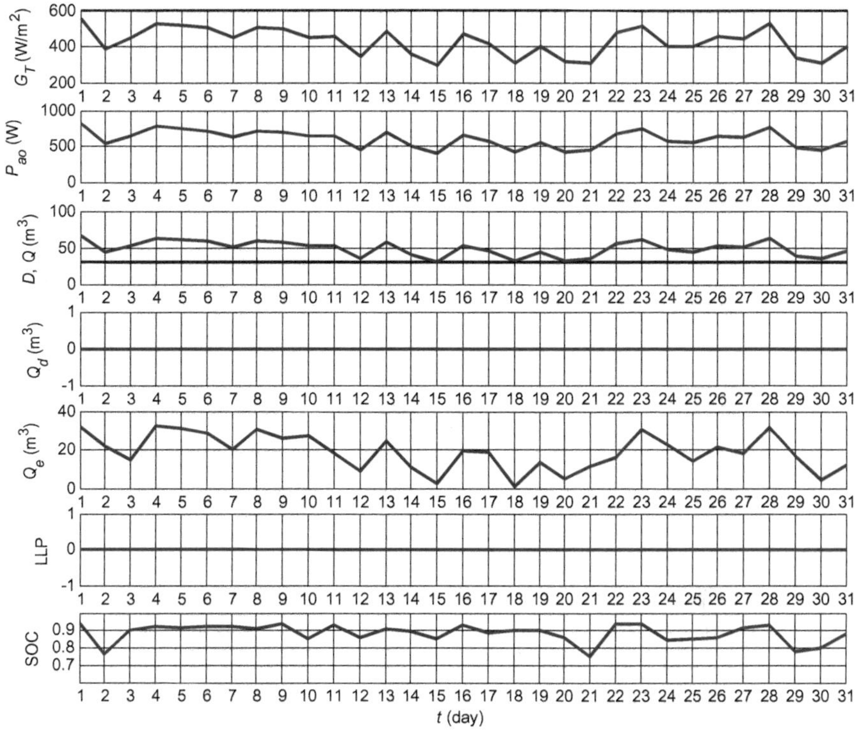

Fig. 4.12 Performance of PVPS based on optimal configuration in March.

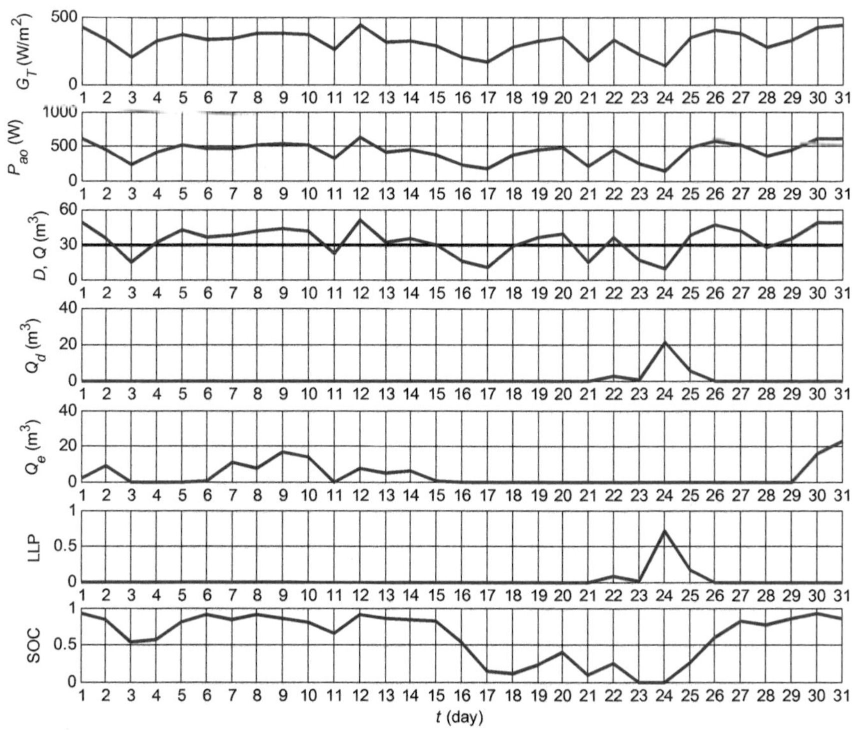

Fig. 4.13 Performance of PVPS based on optimal configuration in December.

the value of the water demand per day in December. The lowest average daily pumped water of the system is on December 24 at around 8.732 m^3. The fourth and sixth parts of the figure explain the shortage days of PVPS in December, where the daily *LLP* and Q_d are equal to zero in most days of December, except December 22–25. The maximum daily *LLP* occurs on December 24 at about 0.709. The fifth part of the figure explains the daily excess water throughout December, when Q_e is equal to zero for 20 days. The last part of the figure shows the daily state of charge of the storage tank. In most days of December, the *SOC* of the storage tank is more than 50%, except December 16–26, when the *SOC* of the tank is less than 40%.

The PV array efficiency is stable at around 0.08 during most of the operation hours and slightly increases to 0.083 at 10 a.m. By contrast, the subsystem and overall efficiencies have the same trend and fluctuate during most hours of the day between the ranges [0.5, 0.736] and [0.041, 0.061], respectively. The maximum efficiencies of the subsystem and overall PVPS are 0.736 and 0.061, respectively, at 1 p.m.

Fig. 4.14 shows the monthly productivity of the PVPS relative to the solar radiation, ambient temperature, and water demand throughout 1 year.

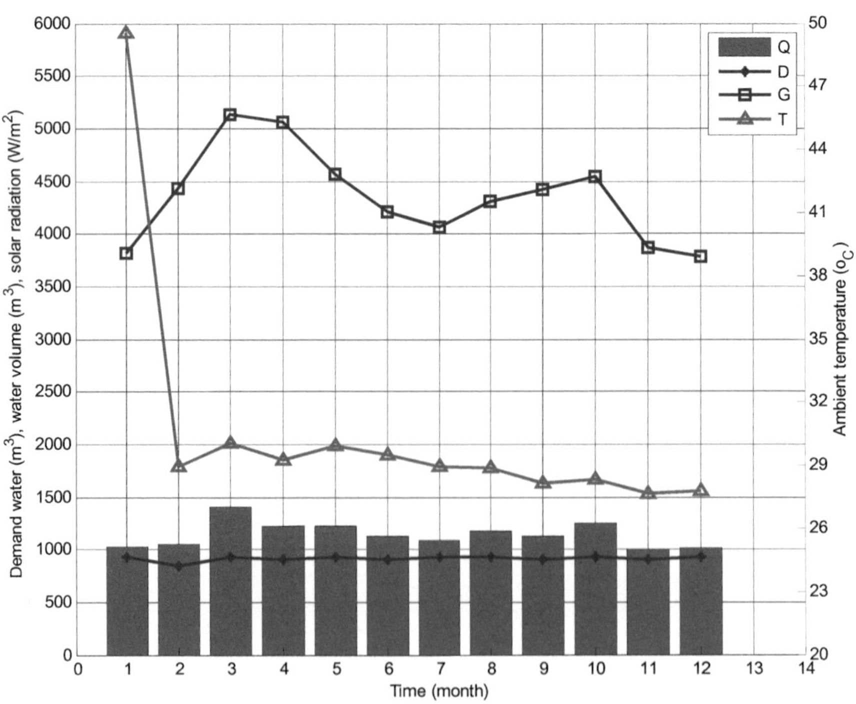

Fig. 4.14 Monthly solar radiation, ambient temperature, water demand, and water production of PVPS throughout a year.

Table 4.12 Performance data of the system on shortage days.

Seq.	Date	G_T (W/m^2)	Q (m^3/day)	Q_d (m^3)	*LLP*	*SOC*
1	01/Jan	4663.800	44.354	4.338	0.145	0.374
2	21/Jan	2163.900	11.529	8.656	0.289	0
3	22/ Jan	2783.100	19.792	10.208	0.340	0
4	23/ Jan	3602.900	29.117	6.466	0.216	0.112
5	09/Nov	4197.200	37.524	2.223	0.074	0.229
6	22/Dec	3966.500	35.555	2.510	0.0837	0.250
7	23/Dec	2638.900	17.083	0.443	0.015	0
8	24/Dec	1591.700	8.732	21.269	0.709	0
9	25/Dec	4174.800	38.358	4.904	0.164	0.265

The maximum productivity ofPVPS occurs in March at about 1407.505 m^3. By contrast, the minimum productivity ofthe system occurs in November at about 996.730 m^3 with 3869.177 W/m^2 as the average daily monthly solar radiation. The average monthly ambient temperature values fluctuate between 27.5 and 30°C throughout the year, except in January, when the ambient temperature is 51.526 °C.

In most days of the year, the proposed PVPS satisfies the water demand, and the daily ***LLP*** is zero, except for 9 days: January 1 and 21–23, November 9, and December 22–25. Table 4.12 illustrates the average daily solar radiation, the daily ***LLP***and ***SOC*** of the storage tank, the deficit, and pumping the water for these days. The maximum shortage occurs on December 24 at about 21.269 m^3 with the ***LLP*** equal to 0.709. By contrast, the minimum shortage occurs on December 23 at around 0.443 m^3 with the ***LLP*** equal to 0.015. Although the daily productivity of the system exceeds the daily water demand on January 1, November 9, December 22 and 25, the daily ***LLP*** values of the system are not zero. This can be justified given that the amount of resident water in the storage tank for the aforementioned dates is very little in the early hours of the morning, when it is 1.909 m^3 on December 22 and 0 on January 1, November 9, and December 25.

CHAPTER 5

Artificial intelligence techniques for sizing photovoltaic water pumping system

Contents

5.1 Introduction

Due to the effective role of system configuration in the performance of photovoltaic water pumping systems (PVPS), optimal configuration should be considered when designing the system to meet water demand. Various methodologies have been reported in the literature for sizing renewable energy systems such as intuitive, analytical, numerical iterative (Chapter 4), and artificial intelligence methods. Recently, evolutionary algorithms have been used to size PV systems.

In general, there are two common approaches that are being used to solve a multiobjective problem using evolutionary algorithm. The first approach is called the aggregation method, which aggregates the individual objective functions by a single aggregating function according to the specific weights for each individual objective function. In other words, the multiobjective optimization problem is converted into a single objective

Photovoltaic Water Pumping Systems
https://doi.org/10.1016/B978-0-12-821231-8.00005-6

optimization problem. One of the drawbacks of the aggregation method is that it identifies only one solution. Furthermore, the weight initialization, which is aggregated by the aggregation function, is the main obstacle of the aggregation method. It is worth mentioning that the results of the optimization problem solved by the aggregation method strongly depend on the weights. The second approach simultaneously optimizes the objective functions so that the Pareto optimal solutions are obtained. In the multiobjective optimization problem, a set of optimal solutions that exists in the search space is called a Pareto optimal solution. The solution belongs to the Pareto optimal solution set if it is not possible to improve an objective function without deteriorating at least one another objective function. The set of feasible and nondominated solutions in the search space of the multiobjective optimization problem is called the Pareto optimal solution set. Meanwhile, the objective function values in the objective space that correspond to the Pareto optimal solution set are called the Pareto front. Fig. 5.1 shows an ideal Pareto front of a multiobjective optimization problem. The blue nodes in Fig. 5.1 refer to the nondominated solutions that constitute the Pareto front, which represent the optimal solutions of a problem. Meanwhile, the red nodes refer to solutions that are dominated by other solutions. It is worth

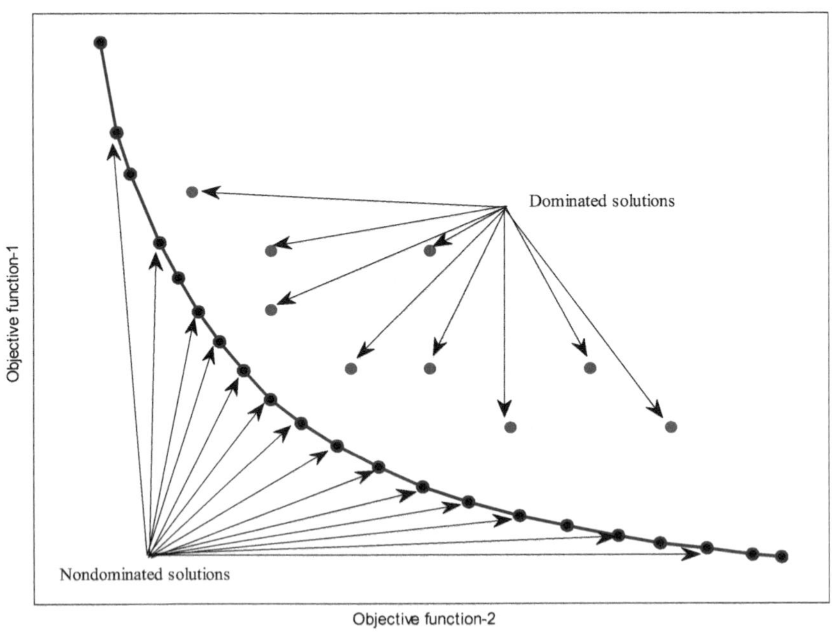

Fig. 5.1 Pareto front of multiobjective optimization problem.

mentioning that the dominated solutions are not represented as optimal solutions for a multiobjective problem. In this chapter, the sizing of the PVPS is considered a multiobjective optimization (MOO) problem.

5.2 Sizing of PVPS based on aggregation of individual objective functions

In this section, the sizing of the PVPS is considered a multiobjective optimization (MOO) problem. The optimal configuration is specified on the basis of minimizing three objectives: loss of load probability, life cycle cost, and excess water. An example of sizing a PVPS based on a multiobjective function evolutionary algorithm using the aggregation concept is presented in the following section.

The best way to understand this method is to describe it using a numerical example. Thus, a conventional differential evolution (DE) algorithm is developed to optimize the size of the PVPS by minimizing an aggregating function that aggregates the weighted loss of load probability (LLP), the life cycle cost (LCC), and excess water (Q_{excess}) as individual objective functions. The modeling of a PVPS is done according to Chapter 3. Meanwhile, the development of the individual objective functions (sizing criteria), the DE algorithm, and the results of this sizing method are discussed here in detail.

The loss of load probability, life cycle cost, and excess water criteria are the most common objectives that are considered to optimize the PVPS size. These criteria are generally contradictory with each other, and therefore a trade-off between the criteria occurs.

The loss of load probability (LLP) is defined as the ratio of the volume of deficit water to the volume of water demand within a period of time. In this research, the LLP is evaluated for 1 year, as shown below

$$\mathrm{LLP} = \frac{\sum_{t=1}^{12*365} Q_d(t)}{\sum_{t=1}^{12*365} D(t)} \tag{5.1}$$

The life cycle cost (LCC) is the most commonly used method to evaluate the cost of a PVPS. The LCC of the PVPS comprises three main parts: the initial capital cost, the present value of the maintenance cost, and the present value of the replacement cost, as expressed in (5.2).

$$LCC = IC + MC + RC, \tag{5.2}$$

where IC is the initial capital cost (USD), MC is the present value of the maintenance cost (USD), and RC is the present value of the replacement cost (USD).

The initial capital cost of a PVPS consists of the price of the system's components, the cost of civil works, and the expenditures for system design and installation. The initial capital cost of a PVPS can be expressed by

$$IC = CA_{PV}*UC_{PV} + CA_C*UC_C + CA_{MP}*UC_{MP} + CA_T*UC_T + ICI, \tag{5.3}$$

where CA_i is the capacity of the ith component of PVPS, UC_i is the cost per unit of the ith component (USD/unit), and ICI is the total constant cost, including the cost of installation and civil works (USD). The unit cost of a system's components are illustrated in Table 5.1.

The present value of the maintenance cost of PVPS is given by,

$$MC_r = \begin{cases} MC_{0r}*\left(\frac{1+FR}{IR-FR}\right)\left(1-\left(\frac{1+FR}{1+IR}\right)^{LP}\right) & \text{if } IR \neq FR \\ MC_{0r}*LP & \text{if } IR = FR \end{cases} \tag{5.4}$$

The maintenance cost of a system's components in the first year can be expressed as a percentage of the initial capital cost of the components:

$$MC_{0r} = k_r*IC_r, \tag{5.5}$$

The total maintenance cost of a PVPS is given by,

$$MC = \sum_r^3 MC_r, \tag{5.6}$$

where FR is the annual inflation rate, IR is the annual interest rate, LP is the lifetime of the PVPS (years), MC_{0r} is the maintenance cost of the rth component in the first year (USD), MC_r is the maintenance cost of the rth

Table 5.1 Financial data of system's components.

Seq.	Component	Cost/unit	Life time (year)	k_r	N_r	LP	IR	FR
1	PV	1 USD/Wp	20	1%	0	20	8%	4%
2	Converter	0.5 USD/W	10	0%	1			
3	Motor-pump set	0.75 USD/W	10	3%	1			
4	Storage tank	20 USD/m^3	20	1%	0			

component (USD), IC_r is the initial cost of the rth component (USD), k_r is a constant that refers to the maintenance cost as a percentage of the initial capital cost of the rth component, and the r values are 1, 2, and 3, which are equivalent to the PV array, motor-pump set, and storage tank, respectively.

The present value of the replacement cost of PVPS is given by

$$RC_k = IC_k{}^* \sum_{j=1}^{N_r} \left(\frac{1+FR}{1+IR}\right)^{\left(\frac{LP*j}{N_r+1}\right)}, \tag{5.7}$$

$$RC = \sum_{k=1}^{2} RC_k, \tag{5.8}$$

where IC_k is the initial cost of the kth component (USD), RC_k is the replacement cost of the kth component (USD), N_r is the number of components replaced over the lifetime of the system, and the k values are 1 and 2, which are equivalent to the converter and motor-pump set, respectively.

The third individual objective function is excess water that is discussed in Chapter 3, specifically Section 3.4.

The DE algorithm consists of four simple and consequent steps: initialization, mutation, crossover, and selection. DE is like other population-based direct search algorithms that use an initial population set (S) that is chosen randomly and comprises candidate solutions. This set consists of N_P individual vectors whereas each vector comprises D_P decision variables that are required to be optimized. The last three steps, mutation, crossover and selection, are repeated per iteration to improve the initial candidate solution until the maximum number of generation G_{max} is reached or the required fitness value is achieved. DE uses N_P D_P -dimensional vectors as a population set (S) to search the optimal decision variables in the search space. The population set is defined as:

$$S^G = \left[X_1^G, X_2^G, \ldots, X_{N_P}^G\right] = \left[X_i^G\right] \tag{5.9}$$

where

$$X_i = \left[X_{1,i}, X_{2,i}, \ldots, X_{D_P,i}\right] = \left[X_{j,i}\right] \tag{5.10}$$

where X_i is the target vector and i is the number of individuals (candidate solutions) of the population ($i=1, 2, \ldots\ldots, N_P$), j is the dimension of the individual vector ($j=1, 2, \ldots\ldots, D_P$), and G is the generation index ($G=1, 2, \ldots, G_{max}$). Fig. 5.2 shows the proposed algorithm used to optimize the size of a PVPS.

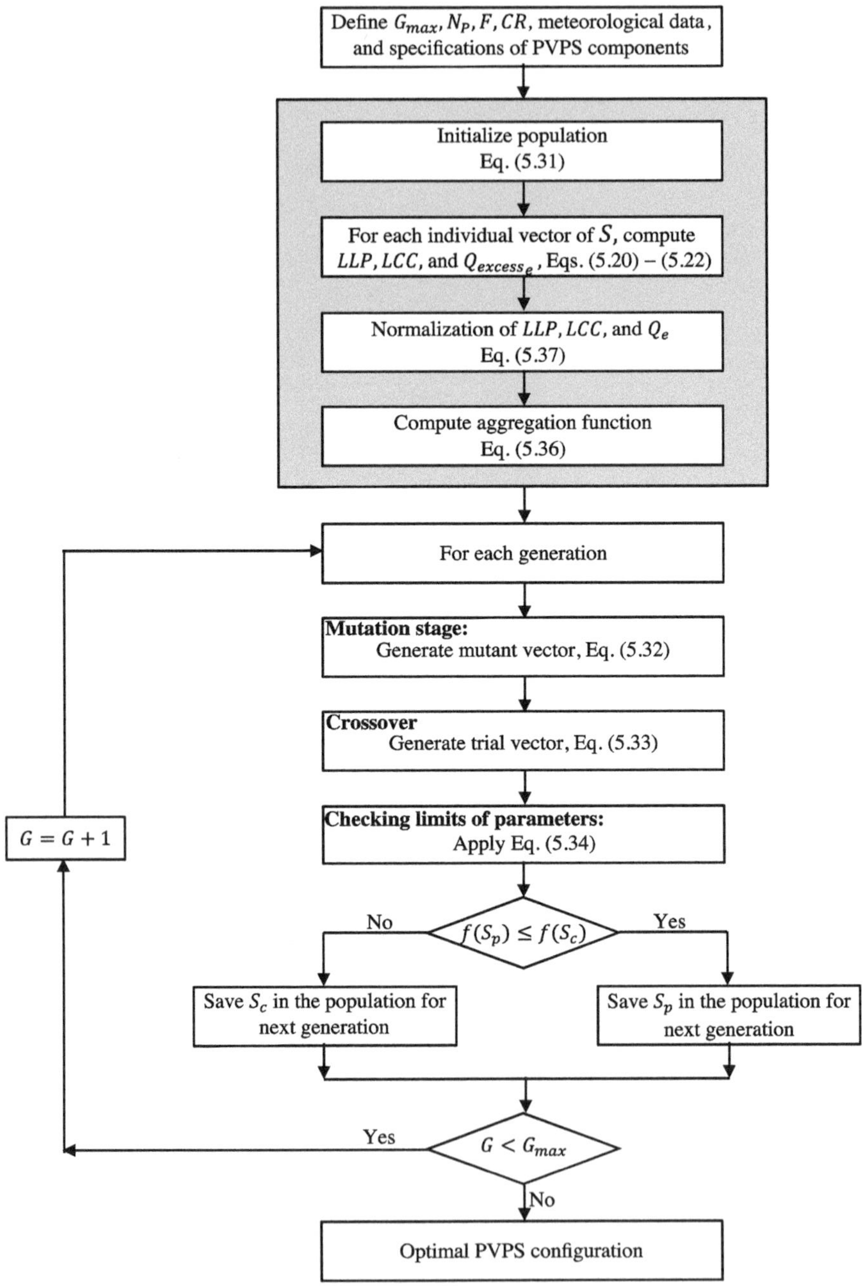

Fig. 5.2 Flow chart of the proposed algorithm.

5.2.1 Initialization

The optimization process begins by creating an initial population; $S^G = [X_i^G]$, $G=0$. The initial values of D_P decision variables are selected randomly and distributed uniformly in the search region. The search region is limited by the lower and upper bounds, which are defined as $X_{j,\ L}$ and $X_{j,\ H}$, respectively. The initial individual vector is selected as;

$$X_{j,i}^0 = X_{j,L,i} + rand\left(X_{j,H,i} - X_{j,L,i}\right) \tag{5.11}$$

where *rand* is a random number within the [0, 1] interval and index *i* refers to the individual vector ($i=1, 2, \ldots, N_P$).

5.2.2 Mutation

DE utilizes a new scheme to generate the trial vector through the addition of the weighted difference between two individual vectors with the third individual vector. For each target vector X_i, a mutant vector $\hat{X}_i^G$ is generated as follows:

$$\hat{X}_i^G = X_\alpha^G + F\left(X_\beta^G - X_\gamma^G\right), \tag{5.12}$$

where X_α^G, X_β^G,and X_γ^G vectors are randomly selected from the population, and α, β, and γ are distinct indices belonging to the range $[1, N_P]$. The vector X_α^G is called the base vector, and F is a mutation scaling control parameter that is typically selected within the range of [0, 1].

5.2.3 Crossover

In this stage, both the target vector (parent solution) X_i^G and the mutant vector $\hat{X}_i^G$ are used to generate a trial vector (candidate solution) $y_{j,\ i}^G$ described as follows:

$$y_{j,i}^G = \begin{cases} \hat{X}_{j,i}^G \ if \ rand \leq CR \ or \ j = I_i \\ X_{j,\ i \ otherwise}^G \end{cases}, \tag{5.13}$$

where *rand* is a random number in the range of (0, 1), I_i is an index that is randomly selected from the range of $[1, D_P]$, and $CR \in [0, 1]$ is the crossover control parameter.

The decision variables of the trial vector should be checked if they are located within the allowed search space to ensure whether the parameter values are physical values. If any parameter is beyond the specified limits of the search space, the parameter will be replaced by a new value as follows:

$$y_{j,i}^G = X_{j,L,i} + rand\left(X_{j,H,i} - X_{j,L,i}\right). \tag{5.14}$$

5.2.4 Selection

The selection stage is applied after the generation of N_P candidate solutions (trial vectors). The selection process between the current parent solution (S_p) and the candidate (S_c) is based on the objective function values for both solutions. The solution that has a small objective function is chosen as a member of the population for the next generation $G+1$. The selection process can be described as:

$$X_i^{G+1} = \begin{cases} y_i^G \text{ if } f\left(y_i^G\right) < f\left(X_i^G\right) \\ X_i^G \text{ otherwise} \end{cases} \tag{5.15}$$

In the current optimization problem, there are three individual objective functions, which are LLP, LCC,and Q_{excess}. These objective functions are aggregated by a single function to deal with the multiobjective optimization problem as a mono-objective problem, as expressed in (5.16).

$$f(x) = \sum_{r=1}^{R} w_r f_r(x), \tag{5.16}$$

where x is the vector of the decision variables that belong to the search space and R refers to the number of individual objective functions. $f_r(x)$ is the rth individual objective function and it is an aggregated function. w_r is the weight of the rth individual objective function $(\sum_{r=1}^{R} w_r = 1)$.

It is worth mentioning that the individual objective functions are not in the same scale. Therefore, a normalization of the objective functions is necessary and it is implemented in this research as below,

$$f_r(x) = \frac{f_r(x) - f_r^{min}(x)}{f_r^{max}(x) - f_r^{min}(x)} \tag{5.17}$$

where $f_r^{min}(x)$ and $f_r^{max}(x)$ are the lower and upper bounds of the rth individual objective function, respectively. Eventually, the reproduction of the candidate solutions (mutation and crossover) and selection stages are repeated until meeting the predefined terminating conditions.

To illustrate the use of a multiobjective-based DE optimization algorithm for optimizing the size of a PVPS, a design example is conducted. The designed PVPS is assumed to be used to supply drinking and irrigation water for 120 people with $30\,m^3$ as the constant total daily water demand. The annual hourly solar radiation and ambient temperature data of Kuala Lumpur, Malaysia, are utilized.

In this example there are three decision variables: the size of the storage tank (C_n) and the number of PV modules that are connected in series (N_s) and parallel (N_p), respectively. These decision variables are optimized to minimize the aggregating function that aggregates three individual objective functions. The search space of the decision variables is [4, 70] for both N_s and N_p and [35, 160] m^3 for C_n. The number of individual vectors in the population is assumed to be 10 D_P. Meanwhile, after conducting many attempts in the implementation of the optimization program, the maximum number of generations is chosen to be 50 to show the stability of the solution. According to Vitaliy, the recommended values of the mutation factor and crossover rate are 0.85 and 0.5, respectively. In the current work, the crossover rate is chosen as 0.5, but the mutation factor is set to be 0.75 instead of 0.85 to improve the performance of the proposed algorithm in terms of convergence to the optimal solution. The weight initialization, which is aggregated by the aggregation function, is the main obstacle of the proposed algorithm. Furthermore, the results of the optimization problem strongly depend on the weights that are chosen by the aggregation function. Therefore, to overcome the complexity of the process of initializing the weights, a wide range of weight sets is tested in this work. The w_1, w_2,and w_3 values are chosen from the effective design spaces [0.3, 0.9], [0.05, 0.55], and [0.05, 0.55] with step size 0.05. Table 5.2 illustrates the weight sets and their results, which achieve acceptable *LLP* values ($LLP \leq 1\%$). The N_s and N_p values are constant within the table and should be equal to 5 and 4, respectively, with at least *a* C_n of 39 m^3. The *LLP* values are inversely related to the w_2 values while the *LCC* values directly correlate to the w_2 values. The correlation between *LLP* and *LCC* is an inverse relationship according to the results in Table 5.2. According to Table 5.2, when the size of a storage tank decreases (considering *N* is constant), the value of *LLP* increases.

The development of *LLP* with the generation of the DE algorithm is shown in Fig. 5.3, when w_1, w_2, and w_3 are 0.6, 0.1, and 0.3, respectively. The *LLP* value is significantly increased to be 0.005024 in the first 18 generations. Fig. 5.4 explains the evolution of *LCC* within the generation evolution of the proposed algorithm. The *LCC* is significantly decreased to about 9911 USD in generation 18. The evolution of Q_{excess} within the generation of DE is illustrated in Fig. 5.5. The Q_{excess} value is dramatically decreased to reach 3500 m^3 in the first 18 generations. Fig. 5.6 shows the development of the aggregation function with the generation evolution of the proposed algorithm to optimize the size of the PVPS. The aggregation function decreased from 0.25 to about 0.05 within the 18 generations.

Table 5.2 Various weight sets and the system's optimal configuration and performance.

w_1	w_2	w_3	LLP	LCC	Q_{excess}	N	N_s	N_p	C_n
0.9	0.05	0.05	0.000682	10,456.673309	3428.531643	20	5	4	76
0.8	0.05	0.15	0.000682	10,456.673309	3428.531643	20	5	4	76
0.7	0.05	0.25	0.000682	10,456.673309	3428.531643	20	5	4	76
0.45	0.05	0.5	0.000682	10,456.673309	3428.531643	20	5	4	76
0.4	0.05	0.55	0.000803	10,433.917837	3430.860544	20	5	4	75
0.8	0.1	0.1	0.005024	9910.541981	3500.083728	20	5	4	52
0.7	0.1	0.2	0.005024	9910.541981	3500.083728	20	5	4	52
0.6	0.1	0.3	0.005024	9910.541981	3500.083728	20	5	4	52
0.5	0.1	0.4	0.005024	9910.541981	3500.083728	20	5	4	52
0.45	0.1	0.45	0.005024	9910.541981	3500.083728	20	5	4	52
0.8	0.15	0.05	0.005572	9865.031037	3508.083728	20	5	4	50
0.4	0.1	0.5	0.005572	9865.031037	3508.083728	20	5	4	50
0.75	0.15	0.1	0.005890	9842.275565	3512.566119	20	5	4	49
0.7	0.15	0.15	0.005890	9842.275565	3512.566119	20	5	4	49
0.65	0.15	0.2	0.005890	9842.275565	3512.566119	20	5	4	49
0.6	0.15	0.25	0.008082	9705.742733	3542.566119	20	5	4	43
0.5	0.15	0.35	0.008082	9705.742733	3542.566119	20	5	4	43
0.75	0.2	0.05	0.008503	9682.987262	3548.176786	20	5	4	42
0.45	0.15	0.4	0.008503	9682.987262	3548.176786	20	5	4	42
0.4	0.15	0.45	0.008503	9682.987262	3548.176786	20	5	4	42
0.7	0.2	0.1	0.009873	9614.720846	3566.176786	20	5	4	39
0.65	0.2	0.15	0.010351	9591.965374	3572.412544	20	5	4	38

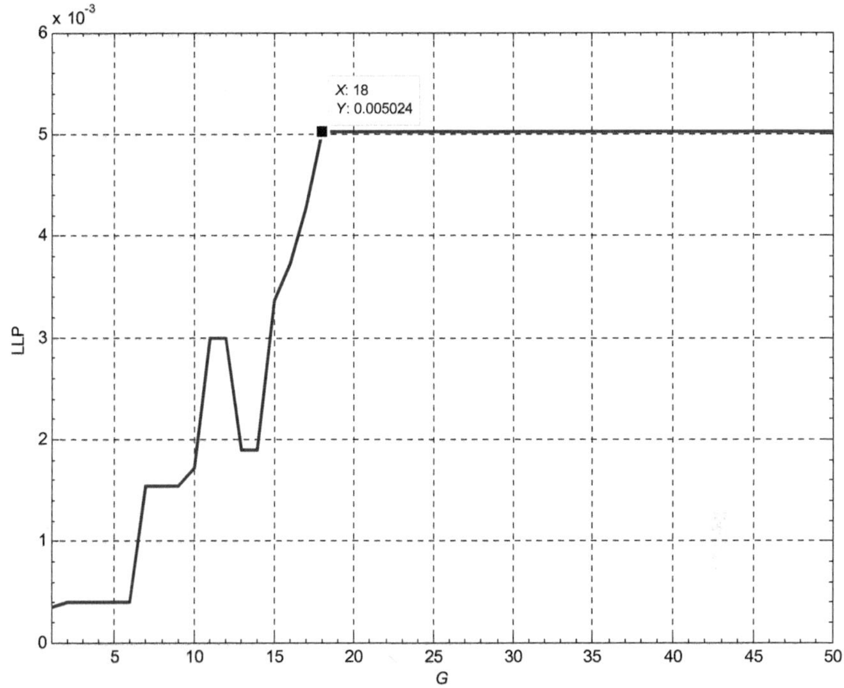

Fig. 5.3 Evolution of loss of load probability (LLP) values.

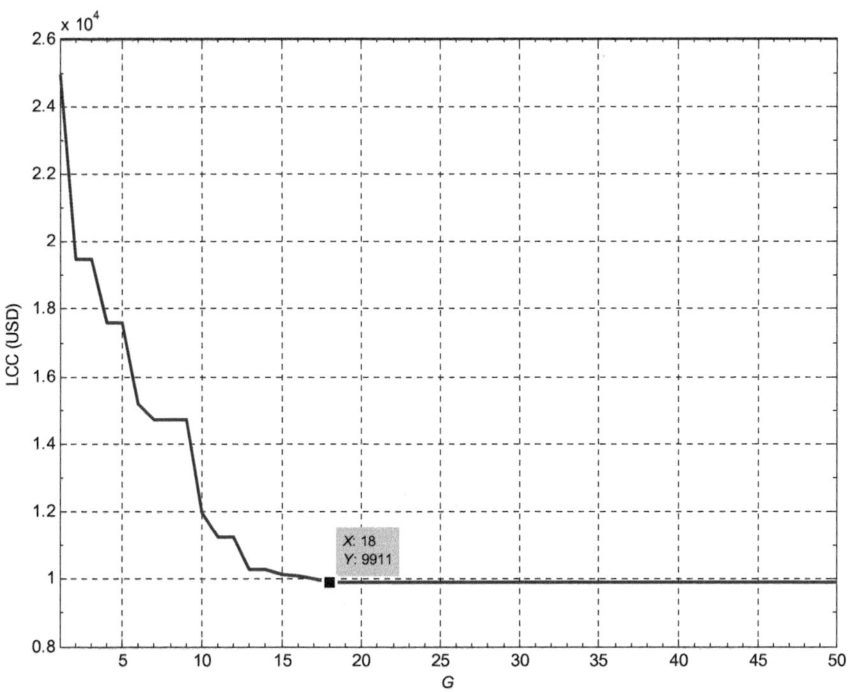

Fig. 5.4 Evolution of life cycle cost (LCC) values.

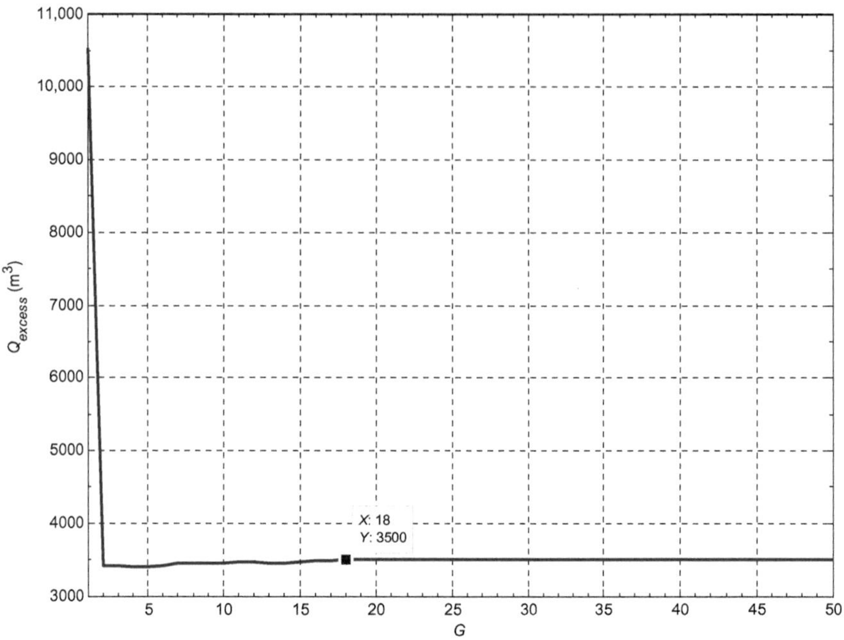

Fig. 5.5 Evolution of excess water (Q_{excess}) values.

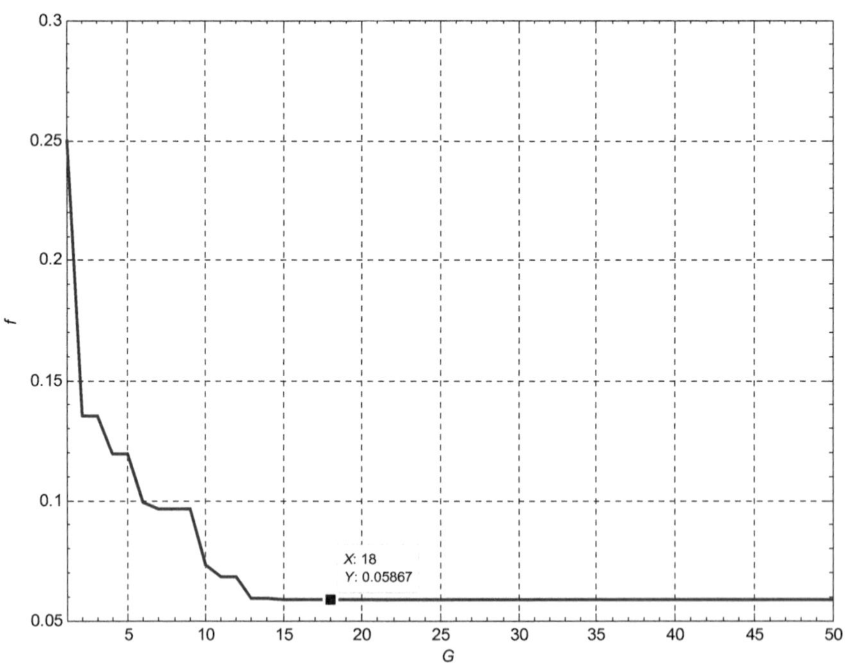

Fig. 5.6 Evolution of aggregating function (f) values.

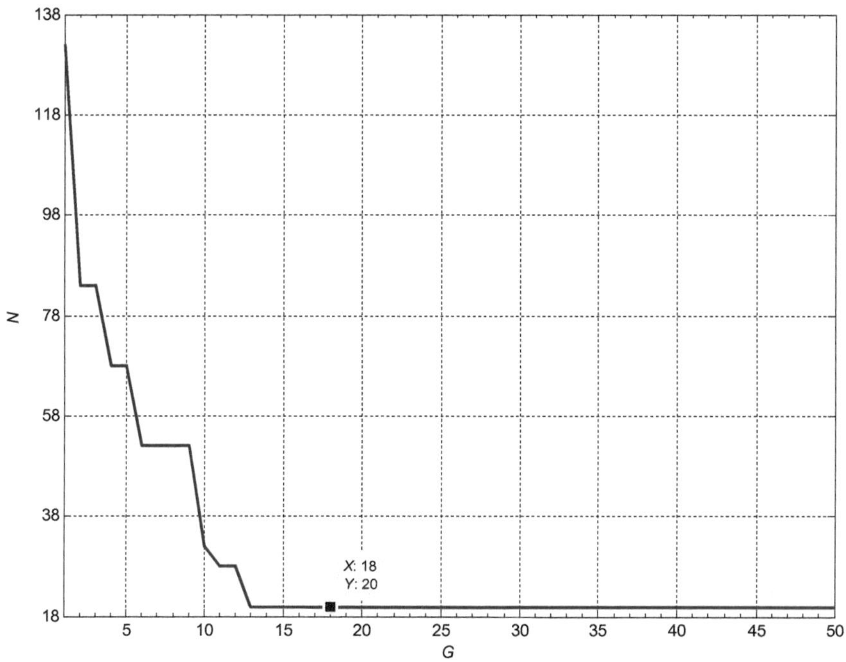

Fig. 5.7 Evolution of number of PV module (*N*) values.

Fig. 5.7 shows the correlation between the number of PV modules and the evolution of generation.

The w_1, w_2, and w_3 are chosen as 0.6, 0.1, and 0.3, respectively, to obtain an *LLP* of 0.5% and to achieve a balance between the *LCC* and Q_{excess}. It is worth mentioning that the proposed method is faster than the numerical sizing method, where the CPU execution time for the proposed and numerical sizing methods are 298.36751 s and 2.5648e3 s, respectively. The numerical sizing method needs a long execution time because it explores all possible PVPS configurations that belong to the design space. Furthermore, the execution time of the numerical sizing method depends on the design space of the PVPS configuration rather than the proposed sizing algorithm, which depends on the number of generations and the mutation factor value.

However, the performance of the system is analyzed based on $N_s=5$, $N_p=4$, and $C_n=52\,m^3$ as an optimal PVPS configuration. The annual hourly performance of the PVPS under optimal configuration is illustrated in Fig. 5.8. The average output power of the PV array over a year is 497.604 W. The average hourly water flow rate over a year is 3.297 m^3/h. The total annual

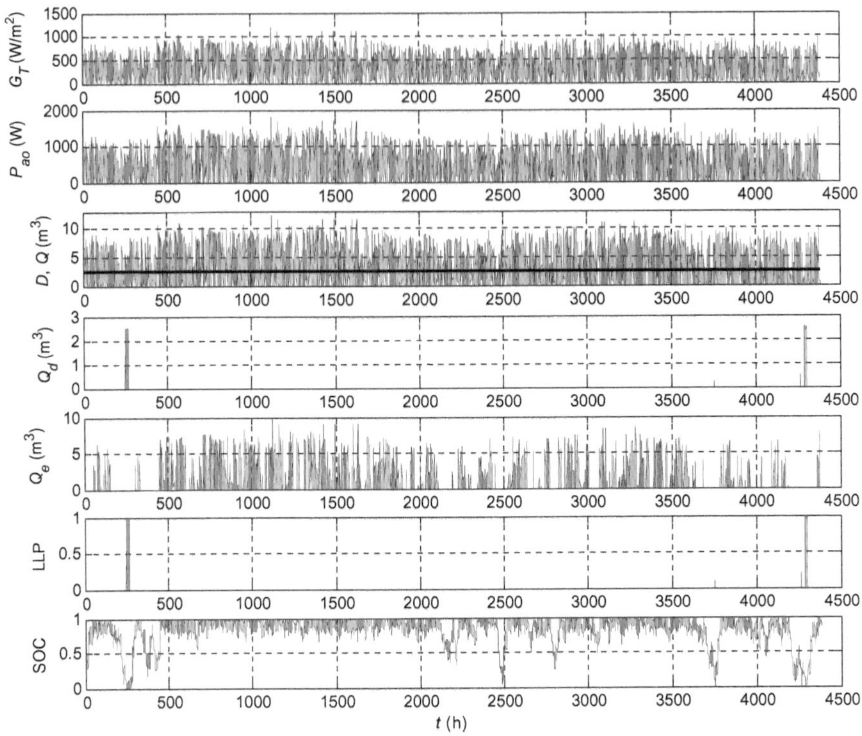

Fig. 5.8 Annual hourly performance of PVPS based on the optimal configuration.

water deficit is 55.015 m^3/year. Finally, the level of water in the storage tank is often greater than 50% over the year, as shown in Fig. 5.9.

The proposed PVPS met the water demand most of the days of the year, except for 9 days, as shown in Fig. 5.10. The maximum shortage occurred on December 24. On the other hand, the minimum shortage occurred on November 9.

Fig. 5.10 also shows the average daily solar radiation and the volume of produced water throughout the shortage days. It is clear that the trend of water production is similar to the solar radiation curve. The productivity of PVPS on November 9 was 37.524 m^3 at 4197.2 W/m^2 as the average daily solar radiation. The productivity on December 24 was 8.732 m^3 with 1591.7 W/m^2 as the average daily solar radiation.

It is worth mentioning that the hourly solar radiation as well as the amount of resident water in the storage tank were very little on the mornings of January 1, November 9, and December 22 and 25. Therefore, the daily *LLP* values for the aforementioned dates were not zero.

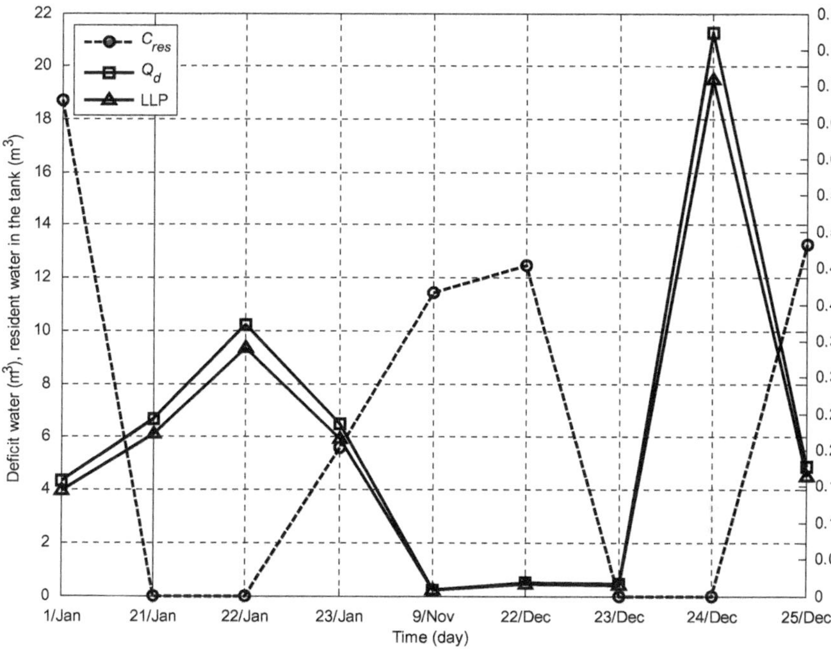

Fig. 5.9 Daily deficit water, current resident water, and LLP during shortage days throughout a year.

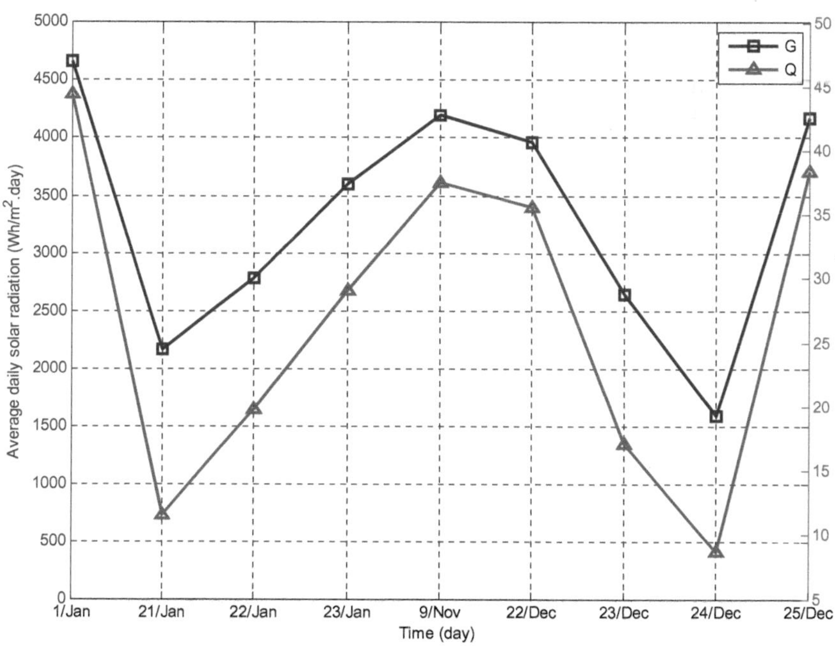

Fig. 5.10 Average daily solar radiation and daily water flow rate during shortage days.

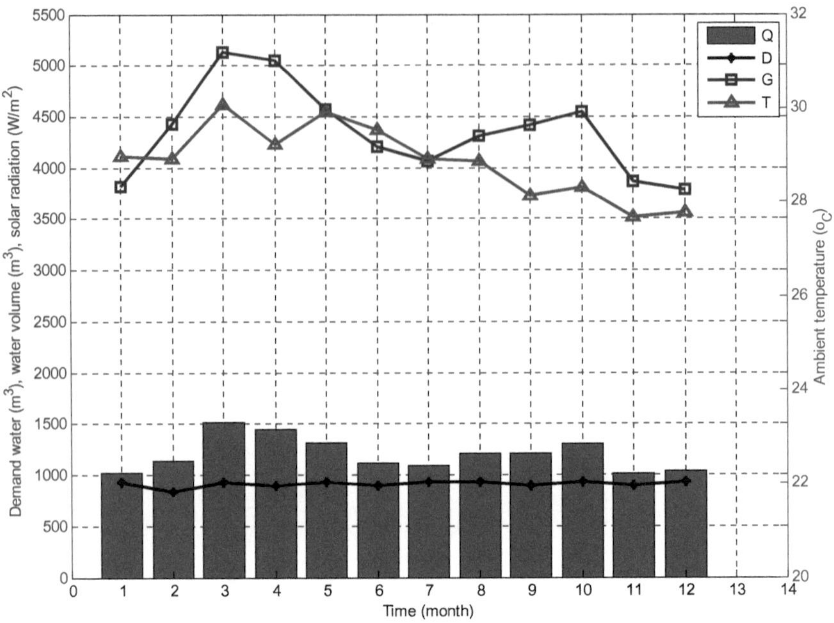

Fig. 5.11 Monthly performance of PVPS based on the optimal configuration.

Fig. 5.11 shows the monthly water flow rate corresponding to the monthly water demand, solar radiation, and ambient temperature. The maximum monthly water production was 1517.553 m^3 in March. On the contrary, the minimum production of the proposed PVPS occurred in November with around 1016.310 m^3.

The following MATLAB code can be used to execute this example:

```
%%%% for optimizing the 3-parameters (Ns, Np & Cn)
%%%% Date 28/11/2019 - Saturday
clc; clear all; close all;
t1=cputime;
% Reading the Solar radiation, Cell temperature, Module voltage and Module
current data %
G=xlsread('Result_Sizing_Aggregating_Function_Method.xlsx',1,'I10:I4389');
%Reading the hourly solar radiation (W) (Year/10-4389),(Month/10-381),
   (Day/10-21)
Tc=xlsread('Result_Sizing_Aggregating_Function_Method.xlsx',1,'J10:
   J4389');
%Reading the hourly cell temperature (K)
Vm=xlsread('Result_Sizing_Aggregating_Function_Method.xlsx',1,'L10:L4389');
%Reading the hourly voltage of one module (V)
```

—cont'd

```
Im=xlsread('Result_Sizing_Aggregating_Function_Method.xlsx',1,'M10:M4389');
%Reading the hourly current of one module (A)
save ('input_variables_of_fitness_function', 'G','Tc','Im','Vm');
%Save G, Tc, Im & Vm data in MAT file

%%%%%%%%%%%%%%%%%//// Control parameter declaration ////%%%%%%%%%%%%%%%%%
K=12;
D=3;                %Dimension of problem (3-parameters of PVPS
Configuration)
NP=10*D;            %///&&&Size of pobulation (number of
individuals)///&&&
Epsilon2=0.23;      %///&&&First Epsilon parameter///&&&
%%%%%%%%%%%%%%%%%%Maximum number of generation (iteration)%%%%%%%%%%%%%%
GEN_max=50;         %///&&&Maximum generation number///&&&
%%%%%%%%%%%%%%%%%%%%%%%%%%Weights Declaration%%%%%%%%%%%%%%%%%%%%%%%%%%
w1=0.6;               %Weight for first f.f. (LLP)
w2=0.1;               %Weight for second f.f. (LCC)
w3=0.3;               %Weight for third f.f. (Qexcess)
%%%%%%%%%%%%%%%%%%//// SAVING RESULTS IN EXCEL FILE ////%%%%%%%%%%%%%%%%%%%
xlswrite ('Result_Sizing_Aggregating_Function_Method.xlsx',NP,1,
         'G4405');
xlswrite ('Result_Sizing_Aggregating_Function_Method.xlsx',GEN_max,1,
         'G4406');
xlswrite ('Result_Sizing_Aggregating_Function_Method.xlsx',
         Epsilon2,1,'G4409');
xlswrite ('Result_Sizing_Aggregating_Function_Method.xlsx',w1,1,
         'G4410');
xlswrite ('Result_Sizing_Aggregating_Function_Method.xlsx',w2,1,
         'G4411');
xlswrite ('Result_Sizing_Aggregating_Function_Method.xlsx',w3,1,
         'G4412');
%%%//// DEFINING THE SEARCH SPACE OF DECISION VARIABLES (Ns, Np and Cn) ////%%%
Ns_l=5;               %Lower limit of no. of series PV modules (Ns)
Ns_h=70;               %Upper limit of no. of series PV modules (Ns)
Np_l=4;                 %Lower limit of no. of parallel PV modules (Np)
Np_h=70;                 %Upper limit of no. of parallel PV modules (Np)
Cn_l=35;                  %Lower limit of storage tank (Cn)
Cn_h=160;                  %Upper limit of storage tank (Cn)
L=[Ns_l Np_l Cn_l];          %Define lower limit vector of 3-decision-
                              parameters
H=[Ns_h Np_h Cn_h];          %Define upper limit vector of
                              3-decision-parameters
```

Continued

—cont'd

```
%%%%%%%%%%%%%Number of times the algorithm is repeated%%%%%%%%%%%%%%%%
rr=1;
%%%%%%%%%%%%%%%%%%%%%%%%%%%%%%%%%%%%%%%%%%%%%%%%%%%%%%%%%%%%%%%%%%%%%%%%
Ns_average=zeros(1,rr);   %Array for saving the best (Ns) values for rr times
Np_average=zeros(1,rr);    %Array for saving the best (Np) values for rr times
Cn_average=zeros(1,rr);     %Array for saving the best (Cn) values for rr
                             times
f_average=zeros(1,rr);       %Array for saving the best (f ) values for rr times
f1_average=zeros(1,rr);       %Array for saving the best (f1) values for rr
                               times
f2_average=zeros(1,rr);        %Array for saving the best (f2) values for rr
                                times
f3_average=zeros(1,rr);         %Array for saving the best (f3) values for rr
                                 times
for b=1:rr                      %Repeat the algorithm 10 times
    %%%%%%%////Algorithm's variables and matrices declaration////%%%%%%
    x=zeros(D,1);               %Trial vector
    pop=zeros(D,NP);            %Population matrix (target matrix)
    sigma_0=zeros(D,1);         %Array (Column vector) of standard
deviation for each row of the initial population
    sigma_g=zeros(D,1);         %Array (Column vector) of standard
deviation for each row of the population of g-iteration
    Fit=zeros(1,NP);            %Overall fitness function matrix of
                                 the population
    Fit1=zeros(1,NP);           %First fitness function matrix of the
                                 population
    Fit2=zeros(1,NP);           %Second fitness function matrix of the
                                 population
    Fit3=zeros(1,NP);           %Third fitness function matrix of the
                                 population
    r=zeros(3,1);               %Randomly selected indicies for
                                 mutation stage
    %%%%%%%%%%%%%%%%////Initializing the population////%%%%%%%%%%%%%%
    for j=1:NP                  %For all individuals vector
        for i=1:D               %For all variables of individual vector
            xx=L(i)+(H(i)-L(i))*rand(1,1);             %Initializing
the individuals vector
            pop(i,j)=round(xx);
        end
        Ns=pop(1,j);              %Specified the Ns value from
                                   population
        Np=pop(2,j);              %Specified the Np value from
                                   population
```

—cont'd

```
        Cn=pop(3,j);              %Specified the Cn value from
                                   population
        [f,f1,f2,f3]=fitness_function(Ns,Np,Cn,w1,w2,w3);
        %Call the function of computing the fitness functions
        Fit(1,j)=f;               %To save the overall fitness function for
                                   initial population
        Fit1(1,j)=f1;             %To save the first fitness function
                                   for initial population
        Fit2(1,j)=f2;             %To save the second fitness function
                                   for initial population
        Fit3(1,j)=f3;             %To save the third fitness function
                                   for initial population
    end
    [n1 iBest]=min(abs(Fit));
    [nn1 iWorst]=max(abs(Fit));
    sigma_0=std(pop(:,:),0,2);
    %Computing the standard deviation for the rows of the initial population
    dist_1=0;
    for t=1:D
        dist_1=dist_1+(sigma_0(t))^2;
    end
    sigma_00=sqrt(dist_1);
    [d1,Best]=min(abs(Fit));
    Fit111=d1;
    Fit222=Fit111;
    %%%%%%%%%%%%%%%%%%%%%%%%////Optimization////%%%%%%%%%%%%%%%%%%%%%%%%
    ANs=zeros(1,GEN_max);         %Initialize array for (Ns) values
    ANp=zeros(1,GEN_max);         %Initialize array for (Np) values
    ACn=zeros(1,GEN_max);         %Initialize array for (Cn) values
    Af=zeros(1,GEN_max);          %Initialize array for (f ) values
    Af1=zeros(1,GEN_max);         %Initialize array for (f1) values
    Af2=zeros(1,GEN_max);         %Initialize array for (f2) values
    Af3=zeros(1,GEN_max);         %Initialize array for (f3) values
    for g=1:GEN_max               %For each generation (iteration)
        %%%%%%%%%%%%%%%%%%%%Crossover rate should be within[0.4,0.7]%
%%%%%%%%%%%%%%%%%%%%%
        CR=0.6;
        sigma_g=std(pop(:,:),0,2);
        %Computing the standard deviation for the rows of the population of
         g-iteration
```

Continued

—cont'd

```
dist_2=0;
for t=1:D
    dist_2=dist_2+(sigma_g(t))^2;
end
sigma_gg=sqrt(dist_2);
for j=1:NP              %For each individual vector

%% STEP TO GENERATE DONOR (MUTATION) VECTOR (w) BY Md_CONCEPT %%
%% SELECTION THREE RANDOMLY INDICES FOR MUTATION %%
        r(1)=floor(rand*NP)+1;          %First random index
        while r(1)==j                   %To ensure ...
            r(1)=floor(rand*NP)+1;      %r(1)not equal j
        end
        r(2)=floor(rand*NP)+1;          %Second random index
        while (r(2)==j)||(r(2)==r(1))   %To ensure ...
            r(2)=floor(rand*NP)+1;      %r(2)not equal j and r(1)
        end
        r(3)=floor(rand*NP)+1;          %Third random index
        while (r(3)==j)||(r(3)==r(1))||(r(3)==r(2)) %To ensure ...
            r(3)=floor(rand*NP)+1;     %r(1)not equal j and r(1)and r(2)
        end
        %%MUTATION STEPS ACCORDING TO Md-Concept%
        %%Mutation factor%%
        F=0.9;
        w=pop(:,iBest)+F.*(pop(:,r(1))-pop(:,r(2)));
        %To create the mutation (donor) vector
    %%%%%%%%%%%%%%%%%//// Crossover steps ////%%%%%%%%%%%%%%%%%%
    Rnd=floor(rand*D)+1;
    for i=1:D
        if (rand<CR)||(Rnd==i)
            x(i)=w(i);
        else
            x(i)=pop(i,j);
        end
    end
    %%%%%//// Checking the 3-PV-parameters of trial vector ////%%%
    %%%%%%%%%%%//// with the boundary constraints ////%%%%%%%%%%%
    for i=1:D
        if (x(i)<L(i))||(x(i)>H(i))
            x(i)=L(i)+(H(i)-L(i))*rand;
            %x(i)=round(x(i));
        end
    end
```

—cont'd

```
        x=round(x(:,1));
        %%%%%%//// Selection the best individual (either the ////%%%%%%
        %%%%%%%%%%%//// trial or current individual vector)////%%%%%%%%
        Ns=x(1);            %Specified the Ns value trial vector
        Np=x(2);            %Specified the Np value trial vector
        Cn=x(3);            %Specified the Cn value trial vector
        [f,f1,f2,f3]=fitness_function(Ns,Np,Cn,w1,w2,w3);
        %Calculate the fitness functions for trial vector
        if (f<=Fit(1,j))
        %Comparison between fitness functions for trial and target vectors
            pop(:,j)=x;
            %Replace the target individual by trial individual vector
            Fit(1,j)=f;
            %Replace the overall fitness function (f.f) of target
             individual vector with the fitness function of trial one
            Fit1(1,j)=f1;
            %Replace the overall fitness function (f.f) of target
             individual vector with the fitness function of trial one
            Fit2(1,j)=f2;
            %Replace the overall fitness function (f.f) of target
             individual vector with the fitness function of trial one
            Fit3(1,j)=f3;
            %Replace the overall fitness function (f.f) of target
             individual vector with the fitness function of trial one
        end
    end                         %End the loop for each individual vectors
    [n iBest]=min(abs(Fit));
    [nn iWorst]=max(abs(Fit));
    ANs(g)=pop(1,iBest);    %To save the best value of (a) for each generation
    ANp(g)=pop(2,iBest);    %To save the best value of (Rp) for each generation
    ACn(g)=pop(3,iBest);    %To save the best value of (Iph) for each generation
    Af(g)=Fit(iBest);       %To save the value of f
    Af1(g)=Fit1(iBest);     %To save the value of f1
    Af2(g)=Fit2(iBest);     %To save the value of f2
    Af3(g)=Fit3(iBest);     %To save the value of f3
end                         %End the loop for each generation (iteration)
[n3 Ibest]=min(abs(Af ));
Ns_average(b)=ANs(1,Ibest);
Np_average(b)=ANp(1,Ibest);
Cn_average(b)=ACn(1,Ibest);
f_average(b)=Af(1,Ibest);
```

Continued

—cont'd

```
 f1_average(b)=Af1(1,Ibest);
 f2_average(b)=Af2(1,Ibest);
 f3_average(b)=Af3(1,Ibest);
 AF=Af';
 AF1=Af1';
 AF2=Af2';
 AF3=Af3';
 ANS=ANs';
 ANP=ANp';
 ACN=ACn';
 AN=ANS.*ANP;
 xlswrite ('Result_Sizing_Aggregating_Function_Method.xlsx',AN,2,'AC3');
 xlswrite ('Result_Sizing_Aggregating_Function_Method.xlsx',ANS,2,'AD3');
 xlswrite ('Result_Sizing_Aggregating_Function_Method.xlsx',ANP,2,'AE3');
 xlswrite ('Result_Sizing_Aggregating_Function_Method.xlsx',ACN,2,'AF3');
 xlswrite ('Result_Sizing_Aggregating_Function_Method.xlsx',AF1,2,'AG3');
 xlswrite ('Result_Sizing_Aggregating_Function_Method.xlsx',AF2,2,'AH3');
 xlswrite ('Result_Sizing_Aggregating_Function_Method.xlsx',AF3,2,'AI3');
 xlswrite ('Result_Sizing_Aggregating_Function_Method.xlsx',AF,2,'AJ3');
end
%%%%%//// Results of optimization of 3-parameters of PVPS ////%%%%%%
f_bestt=sum(f_average)/rr;      %The overall fitness function of the best
                                 individual
f1_bestt=sum(f1_average)/rr;    %The first fitness function of the best
                                 individual
f2_bestt=sum(f2_average)/rr;    %The second fitness function of the
                                 best individual
f3_bestt=sum(f3_average)/rr;    %The third fitness function of the
                                 best individual
Ns_bestt=sum(Ns_average)/rr;
%The best value of (Ns) for each solar radiation and ambient temperature
Np_bestt=sum(Np_average)/rr;
%The best value of (Np) for each solar radiation and ambient temperature
Cn_bestt=sum(Cn_average)/rr;
%The best value of (Cn) for each solar radiation and ambient temperature
N_bestt=Ns_bestt*Np_bestt;
N_average=Ns_average.*Np_average;
xlswrite ('Result_Sizing_Aggregating_Function_Method.xlsx',N_average',2,'AP2');
%To save the N in excel sheet for rr run
xlswrite ('Result_Sizing_Aggregating_Function_Method.xlsx',Ns_average',2,'AQ2');
%To save the Ns in excel sheet for rr run
xlswrite ('Result_Sizing_Aggregating_Function_Method.xlsx',Np_average',2,'AR2');
```

—cont'd

```
%To save the Np in excel sheet for rr run
xlswrite ('Result_Sizing_Aggregating_Function_Method.xlsx',Cn_average',2,'AS2');
%To save the Cn in excel sheet for rr run
xlswrite ('Result_Sizing_Aggregating_Function_Method.xlsx',f1_average',2,'AT2');
%To save the f1 in excel sheet for rr run
xlswrite ('Result_Sizing_Aggregating_Function_Method.xlsx',f2_average',2,'AU2');
%To save the f2 in excel sheet for rr run
xlswrite ('Result_Sizing_Aggregating_Function_Method.xlsx',f3_average',2,'AV2');
%To save the f3 in excel sheet for rr run
xlswrite ('Result_Sizing_Aggregating_Function_Method.xlsx',f_average',2,'AW2');
%To save the f in excel sheet for rr run
t2=cputime;
e=t2-t1
xlswrite ('Result_Sizing_Aggregating_Function_Method.xlsx',e,2,'AB2');
%To save the CPU execution time in excel sheet for rr run
%%%%%%%%%%%%%%%%%%%%%%To save final optimal values%%%%%%%%%%%%%%%%%%%%%%%%%
xlswrite ('Result_Sizing_Aggregating_Function_Method.xlsx',N_bestt,1,'G4417');
%To save the N in excel sheet for rr run
xlswrite ('Result_Sizing_Aggregating_Function_Method.xlsx',Ns_bestt,1,
   'G4403');
%To save the Ns in excel sheet for rr run
xlswrite ('Result_Sizing_Aggregating_Function_Method.xlsx',Np_bestt,1,
   'G4404');
%To save the Np in excel sheet for rr run
xlswrite ('Result_Sizing_Aggregating_Function_Method.xlsx',Cn_bestt,1,
   'G4400');
%To save the Cn in excel sheet for rr run
xlswrite ('Result_Sizing_Aggregating_Function_Method.xlsx',f1_bestt,1,
   'G4414');
%To save the f1 in excel sheet for rr run
xlswrite ('Result_Sizing_Aggregating_Function_Method.xlsx',f2_bestt,1,
   'G4415');
%To save the f2 in excel sheet for rr run
xlswrite ('Result_Sizing_Aggregating_Function_Method.xlsx',f3_bestt,1,
   'G4416');
%To save the f3 in excel sheet for rr run
xlswrite ('Result_Sizing_Aggregating_Function_Method.xlsx',f_bestt,1,
   'G4413');
%To save the f in excel sheet for rr run
```

Meanwhile, the fitness function of this example can be executed using the following Matlab code

```
%%%% MATLAB Script function for computing the fitness functions of
%%%% DEAM Multi-Objective algorithm-Aggregating-Function
%%%% for optimizing the 3-decision-parameters (Ns, Np and Cn)
%%%% Date 26/11/2019
function [f,f1,f2,f3]=fitness_function(Ns,Np,Cn,w1,w2,w3)
% Define a function to compute the fitness functions
% Cn
% Ns
% Np
%%%%%%%%%%%%%%%%//// Declaration the constants ////%%%%%%%%%%%%%%%%%%%%
load ('input_variables_of_fitness_function', 'G','Tc','Im','Vm');
%Loading G, Tc, Im & Vm data from MAT file
%%%%%%%%%%%%%////COMPUTING FIRST FITNESS FUNCTION (LLP)////%%%%%%%%%%%%%%%%%%
%%%%%%%%%%%%%%%% h1 and h2 are changed according to pumping head %%%%%%%%%%%%%%%
%%h1=20, 30, 40, 50 and 60, respectively.
%%h2=0.1444, 0.1907, 0.2371, 0.2835 and 0.3298, respectively
h1=20;
%%%&&&Factors of head equation we have got it from head calculations
h2=0.1444;
%%%&&&Factors of head equation we have got it from head calculations
%%%%%%%%%%%%%%%%%%%%%%%%%%%%% PV ARRAY %%%%%%%%%%%%%%%%%%%%%%%%%%%%%%%%
Va=Ns.*Vm;                  %Computing the hourly voltage of PV array (V)
Ia=Np.*Im;                  %Computing the hourly current of PV array (A)
Vpv=Va;                     %Output voltage of PV array for storing
                             purpose
Ipv=Ia;                     %Output current of PV array for storing
                             purpose
Pao=Va.*Ia;                 %Computing the hourly output power of PV
                             array (W)
Am=0.9291;                  %The area of one PV module (m^2)
A=Ns*Np*Am;                 %Computing the area of PV array
(m^2)
Pai=A.*G;                   %Computing the hourly input power
                             of PV array (W)
Pai1=(Pai==0).*1;
Pai2=Pai+Pai1;
effa=Pao./Pai2;             %Computing the hourly efficiency
                             of PV array
%%%%%%%%%%%%%%%%%%%%%%%%%%%%% MOTOR %%%%%%%%%%%%%%%%%%%%%%%%%%%%%%%%
```

—cont'd

```
effconv=0.9;                      %DC-DC converter efficiency
Va=0.95.*Va;                      %The output voltage of DC-DC
                                  converter
Ia=0.9.*Ia;                       %The output current of DC-DC
                                  converter
Ra=0.8;                           %Armature resistance of DC motor
                                  (Ohm)
Km=0.175;                         %Torque and back emf constant (V/
                                  (rad/sec))
%TC=0.08;                         %Torque constant for rotational
                                  losses
%VT=0.01;                         %Viscous torque constant for
                                  rotational losses
Ebb=Va-(Ra.*Ia);                  %Computing the hourly back emf
                                  voltage of motor (V)
%%%%%%%% The case of overcurrent supplied to motor by PV array %%%%%%%
Eb=(Ebb>=0).*Ebb;                 %Set Eb=0 when Ebb<0
                                  (overcurrent)/turn off motor
Ia=(Ebb>=0).*Ia;                  %Set Ia=0 when Ebb<0
                                  (overcurrent)/turn off motor
Va=(Ebb>=0).*Va;                  %Set Va=0 when Ebb<0
                                  (overcurrent)/
                                  turn off motor
%%%%%%%%%%%%%%%%%%%%%%%%%%%%%%%%%%%%%%%%%%%%%%%%%%%%%%%%%%%%%%%%%%%%%
Tm=Km.*Ia;                        %Computing the hourly torque of DC motor
Tmm=(Tm==0).*1;
Tm1=Tmm+Tm;
Rou=1000;                   %Density of water (Kg/m^3)
g=9.81;                     %Acceleration due to gravity (m/Sec^2)
d1=33.5*0.001;              %Inlet impeller diameter (mm)
d2=160*0.001;               %outlet impeller diameter (mm)
beta1=38*2*pi/360;          %Inclination angle of impeller blade at
                            impeller inlet (degree)
beta2=33*2*pi/360;          %Inclination angle of impeller blade at
                            impeller outlet (degree)
b1=5.4*0.001;               %Height of impeller blade at impeller inlet (mm)
b2=2.2*0.001;               %Height of impeller blade at impeller
                            outlet (mm)
Kp=Rou*2*pi*b1*(d1/2)^2*tan(beta1)*((d2/2)^2-((b1*(d1/2)^2*tan
   (beta1))/(b2*tan(beta2))));
%Computing the hourly output power of DC motor (W)
```

Continued

—cont'd

```
Pdev=Eb.*Ia;                          %Computing the hourly development power
                                       by motor (W)
Omega=abs(sqrt((Km.*Ia)./Kp));        %Computing the hourly angular speed of
                                       motor (rad/sec)
Pmo=Pdev;
%By neglecting rotational losses, the mecganical power equals to development
power by motor
Pmi=Pao.*0.9;                         %Computing the hourly input power of DC
                                       motor (W)
PMI1=(Pmi==0).*1;                     %To overcome divided by zero
PMI2=Pmi+PMI1;                        %To overcome divided by zero
effm=Pmo./PMI2;                       %Computing the hourly efficiency
                                       of DC motor
%%%%%%%%%%%%%%%%%%%%%%%%%% PUMP %%%%%%%%%%%%%%%%%%%%%%%%%%%%%%
Tp=Tm;      %-TC-0.1.*VT.*Omega;
%The produced torque by motor is equal the torque required for pump (Nm)
Eh=Tp.*Omega;                    %Computing the hydraulic energy (W)
Ppo=Eh;                          %Computing the hourly output power of pump (W)
Ppo=(Ebb>=0).*Ppo;
Ppi=Pmo;                         %Computing the hourly input power of pump (W)
PPI1=(Ppi==0).*1;                %To overcome divided by zero
PPI2=Ppi+PPI1;                   %To overcome divided by zero
effpp=Ppo./PPI2;                 %Computing the hourly efficiency of pump
                                  %effp=(effpp<=0.95).*effpp;
effp=effpp;
Q=zeros(length(Eh),1);
for ii=1:length(Eh)
    r1=h2*2.725;                      %Computing the flow rate of water
    r2=0;                             %Computing the flow rate of water
    r3=h1*2.725;                      %Computing the flow rate of
                                       water
    r4=-Eh(ii);                       %Computing the flow rate of
                                       water
    r=roots([r1 r2 r3 r4]);           %Computing the flow rate of water
    if (imag(r(1))==0 && real(r(1))>0)       %Choosing the real value of
                                              the flow rate of water
        QQQ=real(r(1));
    elseif(imag(r(2))==0&&real(r(2))>0)      %Choosing the real value of
                                              the flow rate of water
        QQQ=real(r(2));
```

—cont'd

```
    elseif (imag(r(3))==0 && real(r(3))>0)      %Choosing the real
                                                 value of the flow
                                                 rate of water
        QQQ=real(r(3));
    else
        QQQ=0; %If all the roots are complex and/or the real part is negative
number or zero
    end
    QQ(ii,1)=QQQ;    %Hourly flow rate (m^3/h)
end
Q1=(QQ==0).*1;        %To overcome divided by zero
Q2=QQ+Q1;             %To overcome divided by zero
H=Eh./(2.725.*Q2);    %Computing the head of pumping water (m)
%%%%%%%%%% OVERALL SYSTEM %%%%%%%%%%%%%%%%%%%%%%%%%%%%%%%%%%%%%%%%%%%%%%%%
effsub=effm.*effp.*effconv;              %Computing hourly subsystem
                                          efficiency
effoverall=effa.*effm.*effp.*effconv;    %Computing hourly overall
                                          efficiency
QQ=(Ebb>=0).*QQ;                         %QQ=(effpp<=0.95).*QQ;
Q=[0;QQ];                                %To add initial case Q=0 for
                                          programming purposes
d=2.5;                                   %Hourly demand water (m^3/h)
Cr=zeros(length(Q),1);                   %To specify the size of current resident
                                          matrix of storage tank
Qexcess_pv=zeros(length(Q),1);           %To specify the size of excess water
                                          matrix
Qexcess_s=zeros(length(Q),1);
SOC=zeros(length(Q),1);
Qdef_pv=zeros(length(Q),1);              %To specify the size of deficit water
                                          matrix (before tank)
Qdeficit_s=zeros(length(Q),1);           %To specify the size of deficit
                                          water matrix (after tank)
X=Q(2:end,1)-d;                          %Difference between the hourly
                                          production and demand water
%%%%%%%%%%%%%%********** Before Tank **********%%%%%%%%%%%%%%
Qdef_pv(2:end,1)=(X<0).*abs(X);          %Computing hourly deficit water
                                          before tank (m^3)
```

Continued

—cont'd

```
%%%%%%%%%%%%%********** After Tank **********%%%%%%%%%%%%%
C=length(Q)-1;
Qexcess_pv(2:end,1)=(X>=0).*abs(X);         %Computing hourly excess
                                             water before and after tank
                                             (m^3)
for i=1:C
    Cr(i+1,1)=((Cr(i,1)+X(i,1))>=0).*abs(Cr(i,1)+X(i,1));
%To compute the hourly current resident water in the tank (m^3)
    SOC(i+1,1)=Cr(i+1,1)/Cn;     %Computing the hourly state of charge
                                  of storage tank
    if SOC(i+1,1)>=1
        SOC(i+1,1)=1;
        Qexcess_s(i+1,1)=Cr(i+1,1)-Cn;
        Cr(i+1,1)=Cn;
    else
        Qexcess_s(i+1,1)=0;
    end
    Qdeficit_s(i+1,1)=((Cr(i,1)+X(i,1))<0).*abs(Cr(i,1)+X(i,1));
%To compute the hourly deficit water (m^3/h) (after tank)
end
Q=Q(2:end,1);
%Final computing of hourly flow rate of water (m^3)
Qexcess_pv=Qexcess_pv(2:end,1);
%Final computing of hourly excess water before and after tank (m^3)
Qexcess_s=Qexcess_s(2:end,1);
%Final computing of hourly excess water after the tank is filled (m^3)
f3=sum(Qexcess_s);              %Third fitness function
Qdeficit_s=Qdeficit_s(2:end,1);%Final computing of hourly deficit water
                                after tank (m^3)
Qdef_pv=Qdef_pv(2:end,1);        %Final computing of hourly deficit water
                                  before tank (m^3)
Cres=Cr(2:end,1);                %Final computing of hourly current
                                  resident water in tank (m^3)
SOC=SOC(2:end,1);                 %Final computing of hourly state of
                                   charge (SOC)
D=zeros(length(Q),1)+d;            %Constructing the matrix of hourly
                                    demand water (m^3)
                                    %LLPh=Qdeficit_s(1:end,1)./D(1:
                                     end,1);
                                     %Computing the hourly LLP
```

—cont'd

```
LLP=sum(Qdeficit_s(1:end,1))/sum(D(1:end,1));
                                        %Computing the LLP of one year
f1=LLP;                                 %First fitness function
%%%%%%%%%////COMPUTING SECOND FITNESS FUNCTION(f2=LCC)////%%%%%%%%%%%%
CAc=800;                %Total capacity of converter required for system
(W)
CAmp=840;                %Total capacity of motor-pump set required
                          for system (W)
UCpv=1;                   %%Unit cost of PV ($/Wp)
UCc=0.5;                   %%Unit cost of converter ($/W)
UCmp=0.75;                  %%Unit cost of motor-pump set ($/W)
UCt=20;                      %%Unit cost of storage tank ($/m^3)
ICI=4000;                     %%Civil and installation works cost ($)
FR=0.04;                       %%Inflation rate
IR=0.08;                        %%Interest rate
LP=20;                           %%Life time of PVPS
Nr=1;          %%Number of replacement times for motor-pump set and
converter
N=Ns*Np;         %Total number of PV modules
%%%%%%% COST SCRIPT %%%%%%%%%%%%%%%%%%
CApv=120*N;       %Total capacity of PV required for system (Wp)
CAt=Cn;            %Total capacity of storage tank required for system
(m^3)
%%%%%%%%%%%%%%%%% COMPUTING INITIAL COST OF PVPS %%%%%%%%%%%%%%%%%%%%
IC=(CApv*UCpv)+(CAc*UCc)+(CAmp*UCmp)+(CAt*UCt)+ICI;  %Initial cost of PVPS($)
%%%%%%%%%%%%%%%%% COMPUTING REPLACEMENT COST OF PVPS %%%%%%%%%%%%%%%%
RCCmp=zeros(1,2);
%Array to accumalate the replacement cost for every replacement of motor-pump
set
RCCc=zeros(1,2);
%Array to accumalate the replacement cost for every replacement of converter
for j=1:Nr
    RRCmp=(CAmp*UCmp)*(((1+FR)/(1+IR))^((LP*j)/(Nr+1)));
%Replacement cost of motor-pump set ($)
    RRCc=(CAc*UCc)*(((1+FR)/(1+IR))^((LP*j)/(Nr+1)));
%Replacement cost of motor-pump set ($)
    RCCmp(1,j)=RRCmp;
%Accumalate the replacement costs for motor-pump set
    RCCc(1,j)=RRCc;
```

Continued

—cont'd

```
%Accumalate the replacement costs for converter
end
RCmp=sum(RCCmp);
%Sum the replacement cost for L replacement motor-pump set ($)
RCc=sum(RCCc);
%Sum the replacement cost for L replacement converter ($)
RC=RCmp+RCc;
%Computing the replacement cost for PVPS ($)
%%%%%%%%%%%%%%%% COMPUTING OPERATION AND MAINTENANCE COST OF PVPS %%%%%%%
MCpv0=0.01*(CApv*UCpv);
%Maintenance and operation cost of PV in the first year ($)
MCmp0=0.03*(CAmp*UCmp);
%Maintenance and operation cost of motor-pump set in the first year ($)
MCt0=0.01*(CAt*UCt);
%Maintenance and operation cost of storage tank in the first year ($)
MCpv=MCpv0*((1+FR)/(IR-FR))*(1-((1+FR)/(1+IR))^LP);
%Maintenance and operation cost of PV along life time of PVPS ($)
MCmp=MCmp0*((1+FR)/(IR-FR))*(1-((1+FR)/(1+IR))^LP);
%Maintenance and operation cost of motor-pump set along life time of PVPS ($)
MCt=MCt0*((1+FR)/(IR-FR))*(1-((1+FR)/(1+IR))^LP);
%Maintenance and operation cost of storage tank along life time of PVPS ($)
MC=MCpv+MCmp+MCt;
%Calculating the maintenance and operation costs of PVPS along life time of PVPS ($)
%%%%%%%%%%%%%%%% COMPUTING LIFE CYCL COST OF PVPS %%%%%%%%%%%%%%%%%%%%
LCC=IC+RC+MC;                  %Computing the LCC of PVPS ($)
f2=LCC;                        %Second fitness function
%%%%%%%%%%%%%%%%%%%%%%Fitness Function Normalization%%%%%%%%%%%%%%%%%
f1min=0.00023;               %Lower limit of first objective function
f1max=0.90245;               %Upper limit of first objective function
f2min=7339.17365;            %lower limit of second objective function
f2max=19194.77455;           %Upper limit of second objective function
f3min=0;                     %lower limit of third objective function
f3max=31068.5531;            %Upper limit of third objective function
f1=(f1-f1min)/(f1max-f1min);   %Normalization first objective function
f2=(f2-f2min)/(f2max-f2min);   %Normalization second objective
                                function
f3=(f3-f3min)/(f3max-f3min);   %Normalization third objective function
%%%%%%%%%%%%Aggregating Function for multiobjectives%%%%%%%%%%%%%%%%%
f=w1*f1+w2*f2+w3*f3;
end
```

5.3 Sizing of PVPS using a multiobjective optimization algorithm-based Pareto dominance concept

In this section, the sizing of the PVPS is considered as a multiobjective optimization (MOO) problem, which is solved based on the optimality Pareto concept to overcome the obstacles of other sizing methods. A set of optimal configurations is specified on the basis of minimizing two objectives, namely, technical (loss of load probability) and economic (life cycle cost) objectives.

In this example a differential evolution for the multiobjective optimization (DEMO) algorithm is developed to optimize the size of the PVPS by minimizing the multiobjective functions simultaneously. The proposed objective functions are composed of technical and economic objectives. Loss of load probability is used as a technical objective, whereas life cycle cost is considered an economic objective. The proposed PVPS is designed to provide a daily water demand of 30 m^3 with a 20 m static head and a drawdown level. The optimal configuration of the system is selected from an optimal Pareto set of configurations to achieve a balance between the reliability, cost, and excess water of the system. In the following sections, the DEMO algorithm and results of the sizing method are discussed in detail. It is worth mentioning that the modeling of the PVPS and objective functions (sizing criteria) are the same used in the previous design example. Thus, they are not discussed for brevity.

5.3.1 Differential evolution for multiobjective optimization (DEMO)

Recently, evolutionary algorithms (EAs) have been commonly used to solve MOO problems. EAs deal with a group of solutions that help in producing an optimal Pareto front with fast convergence and good diversity. DEMO is an evolutionary optimization algorithm that uses the conventional DE algorithm to solve MOO problems. In this work, DEMO is used to optimize the size of the PVPS as a bi-objective real optimization problem. *LLP* and *LCC* are two conflicting objectives and are utilized to optimize the size of the PVPS. In general, single-objective optimization (SOO) and MOO algorithms have two main differences. First, a single solution is available in SOO algorithms. By contrast, a set of optimal solutions is available for MOO algorithms, as denoted by the Pareto front. Second, in SOO, the offspring solutions replace the parent solution when the last one is worse than the offspring solution. On the contrary, in MOO, the decision of replacement is not straightforward as in SOO. The dominance (Pareto optimality) concept is one of the most important strategies used to achieve replacement in the selection stage. However, the first three stages (initialization, mutation, and crossover) of DEMO are similar to those in the conventional DE algorithm discussed in the previous sections. The terms trial vector, candidate solution, and offspring solution are used interchangeably throughout the rest of the chapter as well as the rest of the section.

5.3.1.1 Selection

The selection stage is applied after the generation of N_P offspring solutions. The selection process between the current parent solution (S_p) and the offspring solution (S_o) is based on three steps, namely, creating the empirical population, nondominated sorting, and crowding distance ranking.

Step A: Creating the empirical population

The first step of the selection stage is creating a population called the empirical population (POP_e). The members of the empirical population are selected from both S_p and S_o. S_p is discarded from POP_e if it is dominated by the corresponding S_o (i.e.,$S_o \prec S_p$, where the symbol $\prec$ refers to a domination relationship) and vice versa. Otherwise, both S_p and S_o are represented as a member of POP_e when the parent and offspring solutions are not dominating each other. The size of the empirical population will be between N_P and $2N_P$. Therefore, the empirical population needs to be truncated to N_P solutions to prepare it for the next generation. The truncation is based on the following two steps:

Step B: Nondominated sorting

The empirical population is decomposed into several nondominated front levels. The solutions nondominated by other solutions are ranked 1, and they form the first front F_1. The solutions that are nondominated except by the solutions belonging to F_1 are ranked 2, and they form the second front F_2,and so on. In general, the solutions are ranked as K if they are only dominated by other solutions belonging to the front $F_1 \cup F_2 \cup \ldots . \cup F_{K-1}$. Fig. 5.12 illustrates the nondominated sorting algorithm used for two objective functions, namely, f_1 and f_2. When nondominated sorting is conducted, the new population prepared for the next generation is filled by solutions that belong to different nondominated fronts. The solutions belonging to the nondominated front with rank 1 are selected first to fill the new population, followed by the solutions ranked as 2, 3, and so on. The size of the empirical population is $2N_P$, so not all solutions may be included in the N_P slots of the new population. The solutions that are excluded in the new population are discarded. The solutions belonging to the last allowed rank may be more than the remaining slots in the new population. In this case, a crowding distance ranking concept is used to select the solutions in the least crowded region instead of arbitrarily discarding some solutions to increase the diversity of the solutions.

Step C: Crowding distance ranking

The crowding distance concept proposed in NSGA-II is applied in DEMO to increase the diversity of optimal solutions. The average distance of two solutions on the left and right of a solution i, along each objective, can be used to estimate the density of the solutions surrounding solution i. Based on that, the crowding distance of a particular solution i can be defined as the circumferences of the rectangle with vertices: the left and right neighbor solutions. The solutions with a high crowding distance rank are considered the best solutions because they introduce more diversity in the population. The crowding distance of a solution i along the mth objective function can be computed as follows:

$$cd_i^m = \frac{f_m^{i+1} - f_m^{i-1}}{f_m^{max} - f_m^{min}}, \tag{5.18}$$

Step1: Sort the solutions according to f_1.

Step 2: If two solutions have the same f_1 value, sort them according to f_2.

Step 3: Initialize the rank value, ($r_k = 1$).

Step 4: While all the solutions are not treated do step 5 to step 11.

Step 5: Let S_i be the first non-treated solution.

Step 6: Set the rank of S_i is r_k, ($R_{S_i} = r_k$).

Step 7: For all non-treated solutions S_j do step 8 to step 10.

Step 8: If S_j and S_i have the same f_1 and f_2 values, then $R_{S_j} = r_k$.

Step 9: If $f_2(S_j) < f_2(S_i)$, then $R_{S_j} = r_k$.

Step 10: Set $S_j = S_i$.

Step 11: Increment r_k, ($r_k = r_k + 1$).

Fig. 5.12 Nondominated sorting algorithm for bi-objective optimization problem.

where cd_i^m is the individual crowding distance value for solution i corresponding to the mth objective function. f_m^{i+1} and f_m^{i-1} are the mth objective function values for the $i+1$ and $i-1$ solutions, respectively. f_m^{max} and f_m^{min} are the maximum and minimum values of the mth objective function, respectively. The overall crowding distance of each solution is computed by the addition of the individual crowding distance values corresponding to each objective. Fig. 5.13 illustrates the algorithm used to compute the crowding distance for each solution in set S with multiobjective functions.

Fig. 5.14 shows the flow chart of the proposed DEMO algorithm to optimize the size of the PVPS.

A DEMO algorithm is proposed to optimally size a PVPS. The PVPS is used to supply water for drinking and irrigation purposes for 120 people. The load profile of the PVPS is considered to be constant, with 30 m^3 as the total daily water demand. The hourly meteorological data of Kuala Lumpur, Malaysia, for 1 year are utilized to size and analyze the performance of the PVPS under optimal configuration. The data are collected at the Subang meteorological station in Subung Jaya, Klang Valley (latitude 3.12° north and longitude 101.6° east). The meteorological data were taken from a monitoring system that consists of a solar radiation transmitter of a high-stability silicon photovoltaic detector model WE300 with an accuracy of ±1%, a temperature sensor for the surface of the PV panel model WE710 with an accuracy of ±0.25 °C, an air temperature sensor model WE700 with a range of −50°C to +50°C and accuracy of ±0.1°C, and a current transducer model CTH-050 with an input range of 0–50 A (DC) and an output of 4–20 mA.

Three decision variables need to be optimized in this work: the size of the storage tank (C_n), the number of PV modules, and the PV array configuration (number of PV modules connected in series and parallel, denoted as

Step1: Determine the number of solutions in S and assign it by L.

Step 2: For each objective function f_m, ($m = 1, 2, 3, \ldots\ldots, M$) sort the solutions of S in ascending order of f_m.

Step 3: For $m = 1, 2, 3, \ldots\ldots, M$ assign infinity distance value for boundary solutions (solutions with highest and smallest objective function values), i.e., $cd_1^m = cd_L^m = \infty$.

Step 4: For $m = 1, 2, 3, \ldots\ldots, M$ compute the crowding distance for all other solutions $i = 2$ to $(L - 1)$ using (5.38).

Fig. 5.13 Algorithm for computing crowding distance.

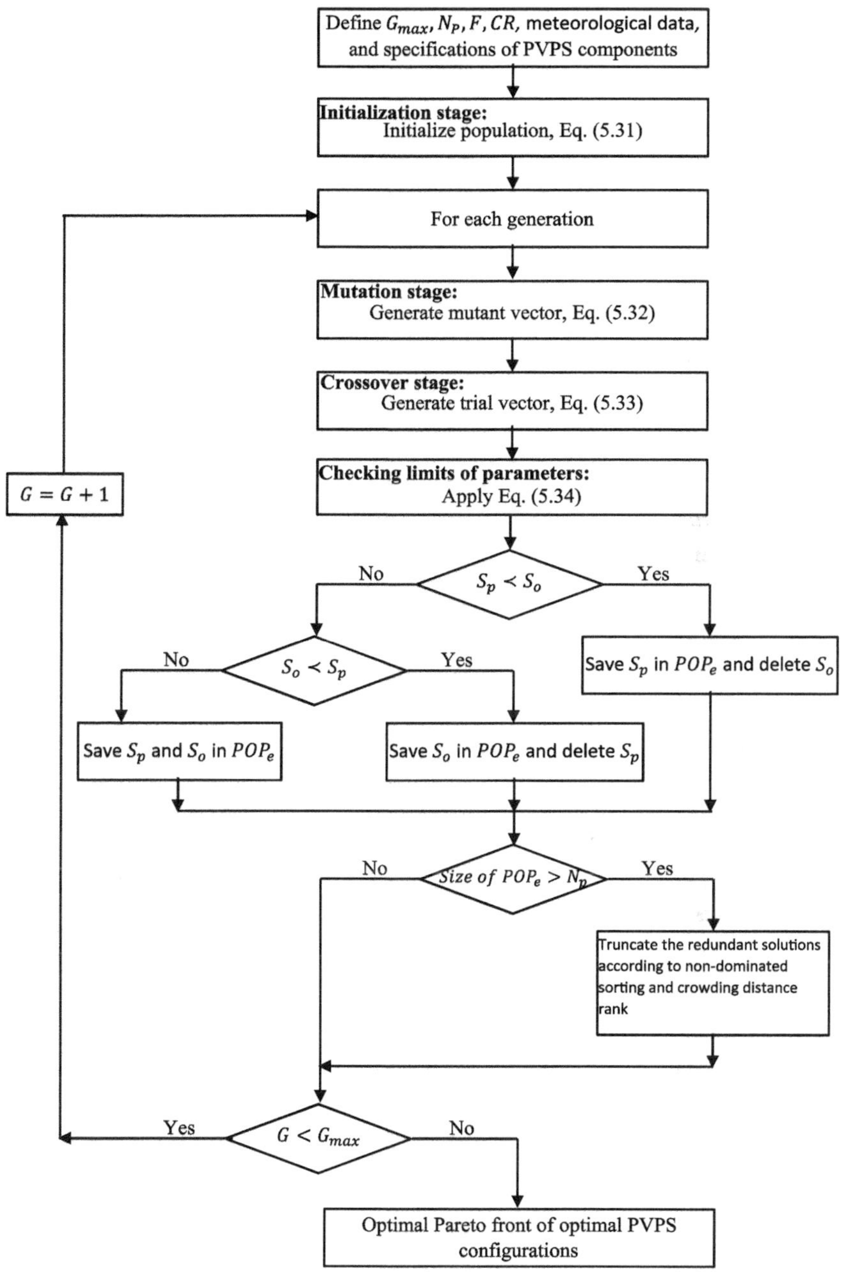

Fig. 5.14 Flow chart of DEMO algorithm.

N_s and N_p, respectively). The search space of the decision variables is [4, 70], [4, 70], and [35, 160] m^3 for N_s, N_p, and C_n, respectively. The mutation factor and the crossover rate are set as 0.75 and 0.5, respectively, in the current work. The population size and the maximum number of generations are assumed to be 50 D_P and 100, respectively. The optimal configurations of the PVPS are obtained with the use of DE algorithm-based bi-objective functions, namely *LLP* and *LCC*. The goal is to derive the optimal configurations that minimize both objective functions. Contrary to the case in SOO methods, a set of optimal solutions called the optimal Pareto front is available in MOO methods because of the trade-off between the conflicting objectives. Fig. 5.15 shows the optimal Pareto front obtained from DEMO. The tradeoff between *LLP* and *LCC* is obvious, as shown in Fig. 5.15. The optimal Pareto front is set within the range [0, 0.01], whereas the desired value of *LLP* is 0.01 or less.

The findings show that the optimal PV array configuration is five PV modules connected in series and four modules connected in parallel to

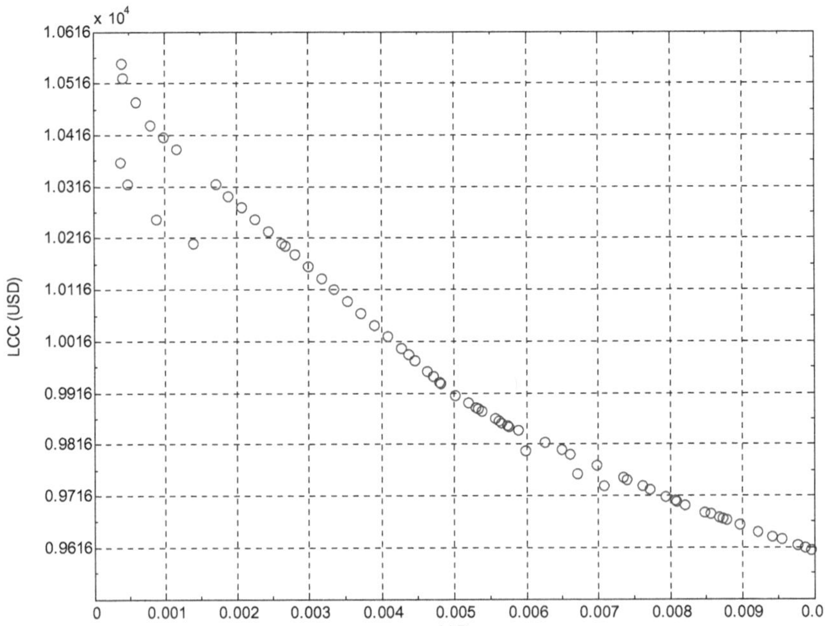

Fig. 5.15 Optimal Pareto front of two objective functions (*LLP* and *LCC*).

achieve an *LLP* value in the range of [0.001, 0.01]. The size of the storage tank is significantly decreased when *LLP* is increased within the range 0.001–0.01. The *LCC* cost of the system per year and the cost of a unit of water are significantly decreased when *LLP* is increased. Furthermore, the excess water is increased when *LLP* is increased. The water deficit is also considerably increased when *LLP* is increased.

Fig. 5.16 shows that the *LCC* of the system is gradually increased when C_n is increased from 35 m^3 to 38 m^3. The correlation between the number of PV modules and *LCC* is depicted in Fig. 5.17. Notably, *LCC* is linearly increased when the number of PV modules (*N*) is increased. When the *N* value is constant and equals 20, the *LCC* of the system is increased from 9614.721 USD to 10,411.162 USD.

The inverse relationship between $LLP \in [0.001, 0.01]$ and C_n in Table 5.5 is confirmed in Fig. 5.18. According to the results of DEMO, the effective size of the storage tank is from 39 m^3 to 80 m^3.

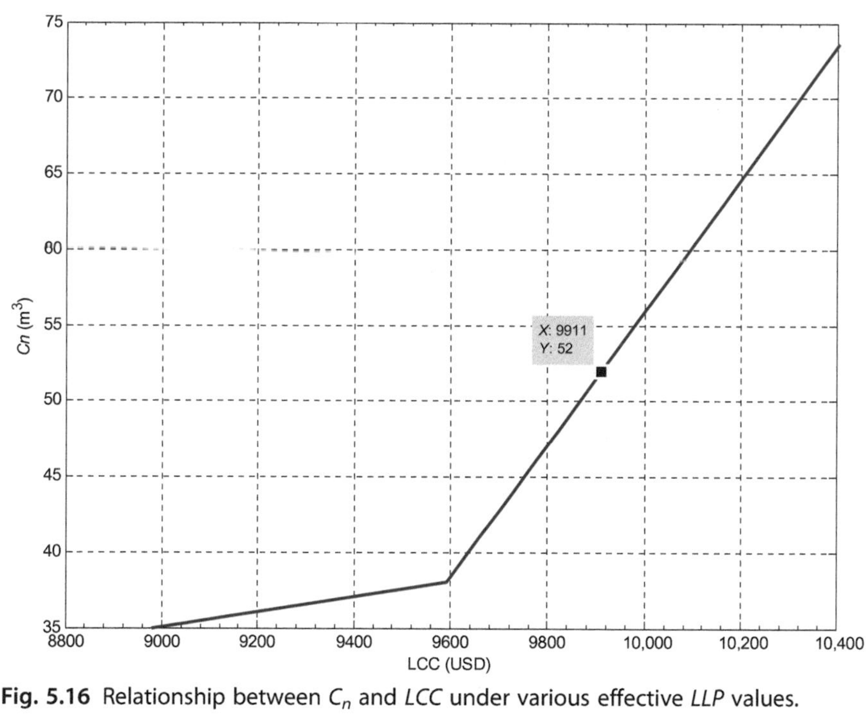

Fig. 5.16 Relationship between C_n and *LCC* under various effective *LLP* values.

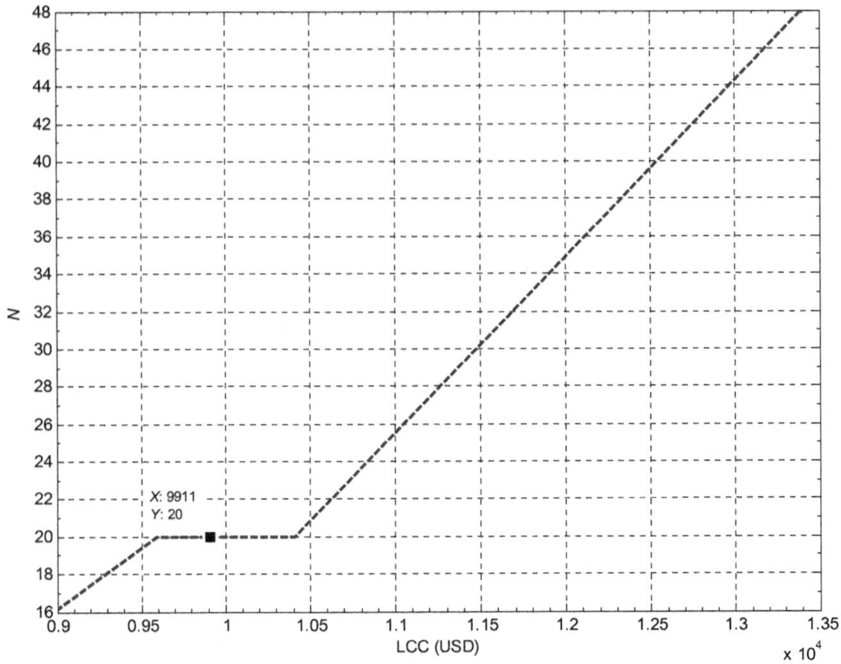

Fig. 5.17 Correlation between *N* and *LCC* under various effective *LLP* values.

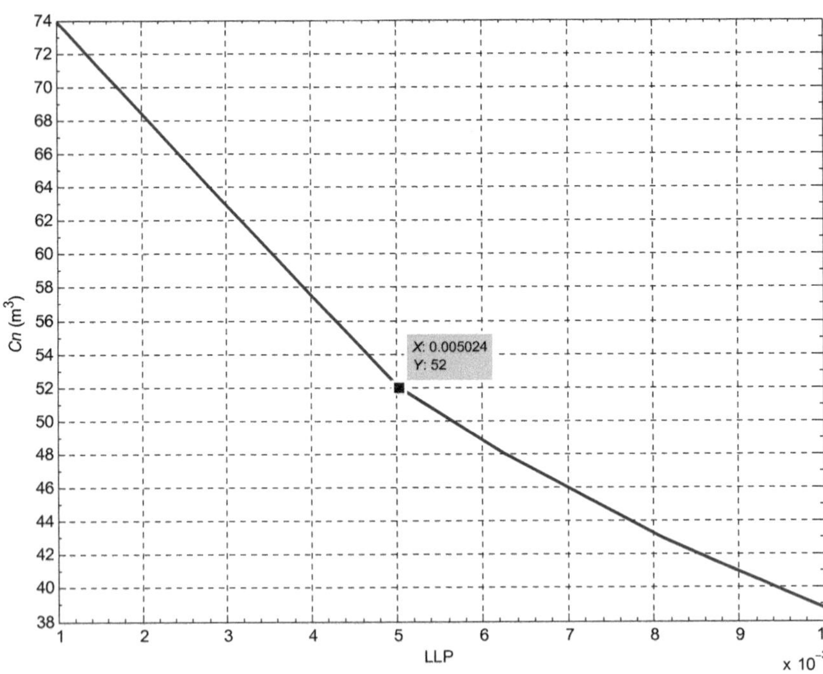

Fig. 5.18 Relationship between C_n and effective values of LLP.

The size of the storage tank is set as 52 m^3. N_s and N_p are selected to be five modules and four modules, respectively, as the optimal configuration to achieve good balance in the *LLP*, *LCC*,excess water, and water deficit.

The performance of the PVPS under the optimal configuration is analyzed, and the daily operation performance of the system over 1 year is presented in Fig. 5.19. The first and second parts of Fig. 5.19 illustrate the daily solar radiation and the daily energy produced by the PV array over 1 year, respectively. The average daily solar radiation and PV array energy production over a year are 4349.859 W/m^2 and 497.604 W, respectively. The third part of Fig. 5.19 indicates that the water flow rate of the PVPS often meets the water demand during most days of the year, except in four periods, as illustrated in the fourth and sixth parts of Fig. 5.19. These parts are related to the water deficit and *LLP*. The proposed PVPS offers high reliability. In this system, the *LLP* value is zero during most days of the year, as shown in the sixth part of Fig. 5.19. The average value of *LLP* over a year is 0.005024. The hourly excess water of the system, accumulated daily over a year, is

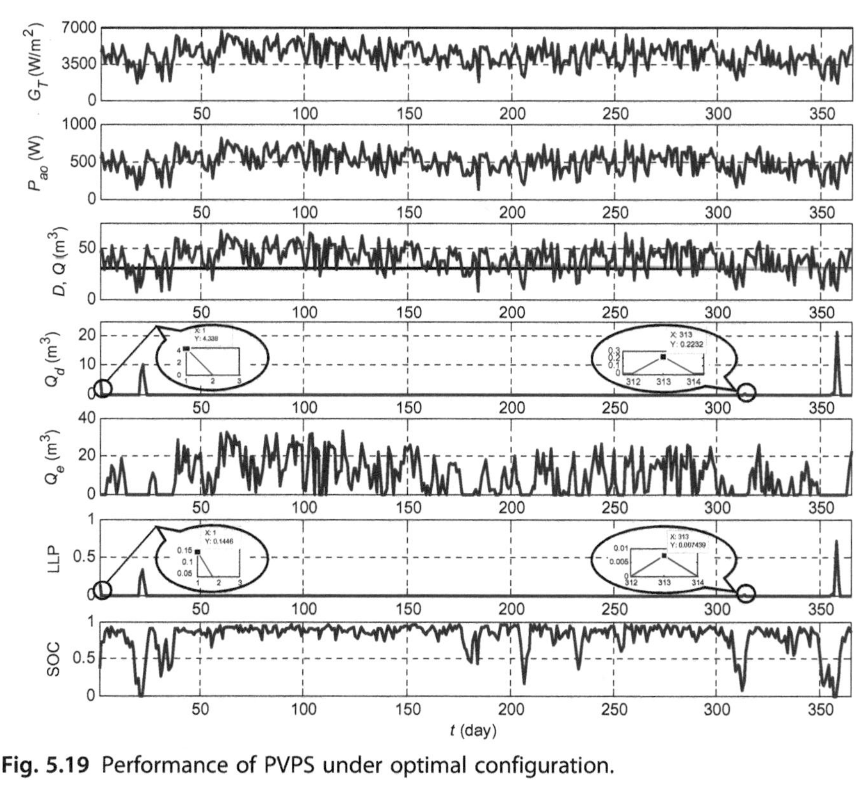

Fig. 5.19 Performance of PVPS under optimal configuration.

depicted in the fifth part of Fig. 5.19. The last part of Fig. 5.19 explains the daily state of charge of the storage tank over a year. The trend of the *SOC* curve indicates that the daily *SOC* is higher than 50% most days of the year because of the high productivity of the system.

The hourly pumping head and the hourly water flow rate during a sunny day (March 1) are illustrated in Fig. 5.20. The average daily solar radiation during this day is 553.233 W/m^2, which is the highest during the entire year. The system starts pumping 1.291 m^3/h with 20.241 m as a head at 8 a.m., with the solar radiation at around 241.7 W/m^2. The productivity of the system is considerably increased to 9.08 m^3/h with a pumping head of 31.906 m at 11 a.m., with the hourly average solar radiation at 847 W/m^2. Then, the productivity of the system is decreased at noon to 8.282 m^3/h with 29.905 m as a head because the solar radiation decreases to 772.2 W/m^2. The maximum productivity of the system is reached at 1 p.m., when the water flow rate, pumping head, and solar radiation are 10.185 m^3/h, 34.980 m, and 958.3 W/m^2, respectively. After 2 p.m., the water flow rate of the PVPS is dramatically decreased to 2.023 m^3/h at 17:00 PM with 20.591 m as a head at 297.2 W/m^2 solar radiation. In the evening (6 p.m.), the productivity of the system is zero, with the solar radiation at 55 W/m^2. Fig. 5.21 shows the

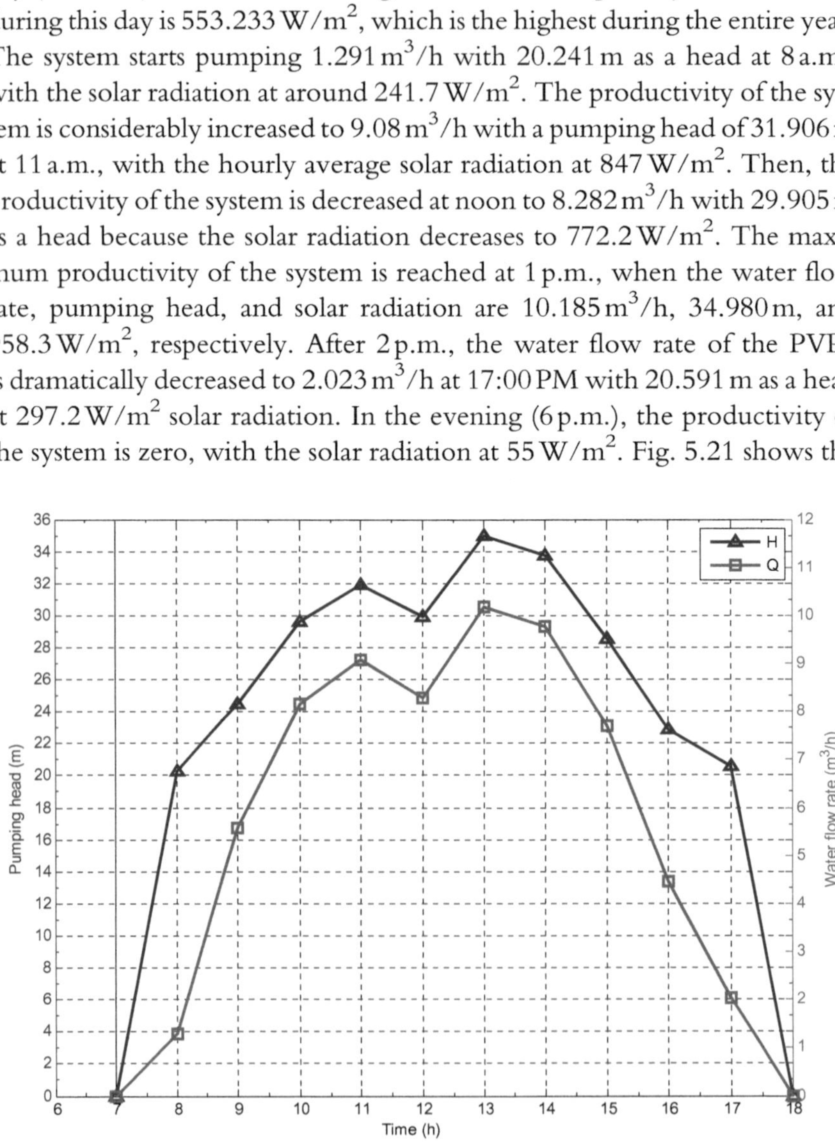

Fig. 5.20 Hourly water flow rate and pumping head during sunny day.

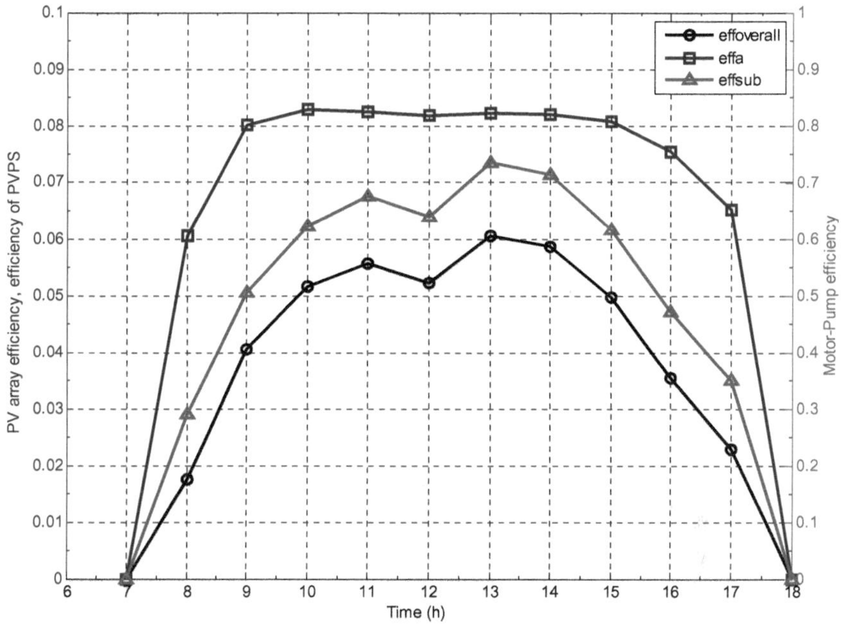

Fig. 5.21 Hourly efficiencies of PV array, subsystem, and overall system during sunny day.

efficiencies of the PV array, subsystem, and overall PVPS on March 1. The PV array efficiency is generally stable during most operation hours, with a value of 0.08. Therefore, the subsystem and overall system efficiencies have the same trend, with average values of nearly 0.47 and 0.04, respectively. The maximum subsystem and overall efficiencies are 0.74 and 0.06, respectively, at 1 p.m.

The hourly productivity of the PVPS during a cloudy day (December 24) is shown in Fig. 5.22. The average daily solar radiation during this day is 132.642 W/m^2. The system starts the operation at noon by producing 0.451 m^3/h with 20.029 m as a head at 163.9 W/m^2 hourly solar radiation. The productivity of the system is significantly increased to reach the maximum value during the day at 3 p.m., when the water flow rate, pumping head, and solar radiation are 3.121 m^3/h, 21.407 m, and 375 W/m^2, respectively. Then, the productivity of the system is sharply decreased to 0.172 m^3/h with 20.004 m as a head at 127.8 W/m^2 solar radiation at 5 p.m. The maximum PV array, subsystem, and overall efficiencies during a cloudy day are 0.07, 0.4, and 0.03, respectively, at 3 p.m., as shown in Fig. 5.23. The average daily efficiencies of the PV array and overall system are less than

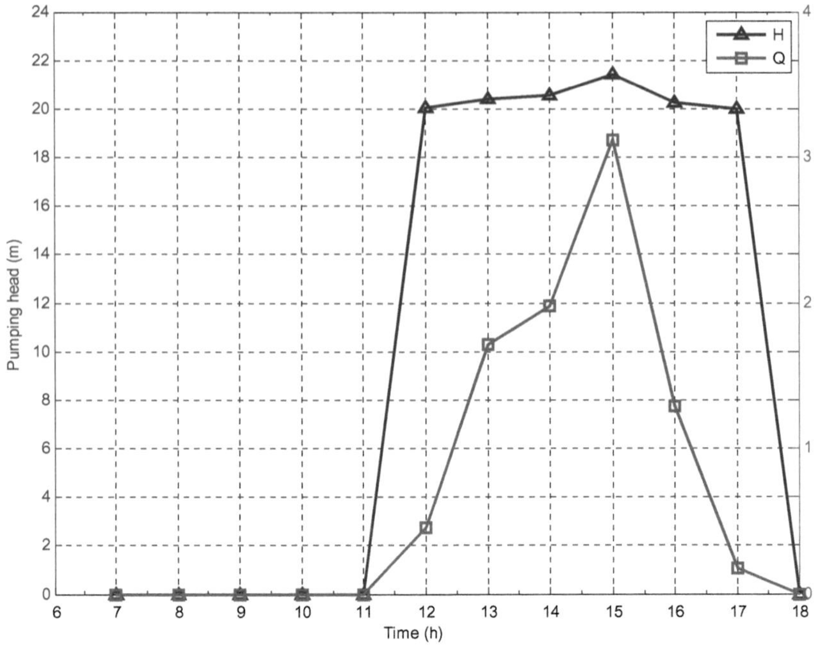

Fig. 5.22 Hourly water flow rate and pumping head during cloudy day.

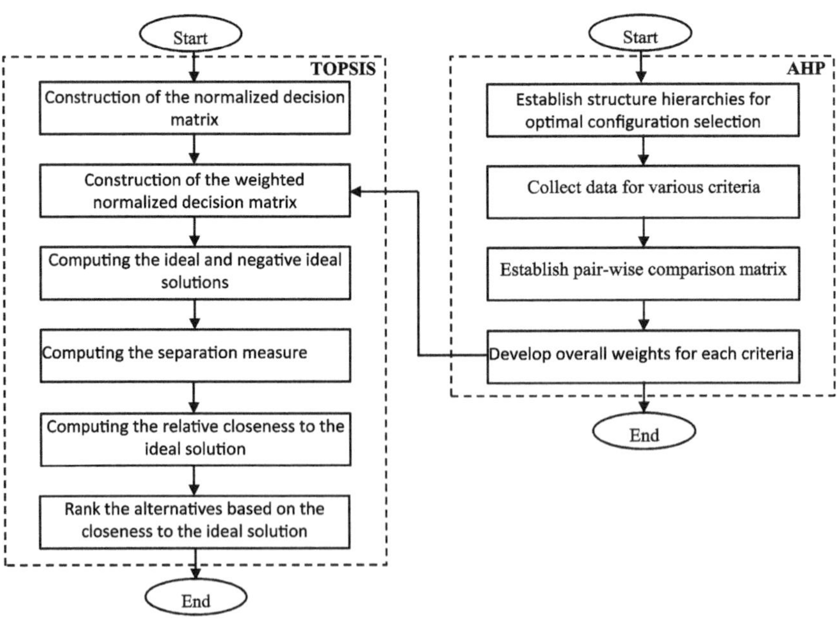

Fig. 5.23 Integrated AHP-TOPSIS model for optimal PVPS configuration selection.

0.06 and 0.02, respectively, and 0.14 for the subsystem. The hourly *LLP* of the PVPS for a cloudy day is 1 from 7 a.m. to 11 a.m., and it is zero only at 3 p.m., as shown in Fig. 5.24. According to results the maximum productivity of the PVPS was reached in March with around 1517.553 m^3 at an average monthly solar radiation of 5131.552 W/m^2. By contrast, the minimum productivity of the system occurred in November with nearly 1016.310 m^3 at 3869.177 W/m^2 as the average daily monthly solar radiation. The average monthly ambient temperature values fluctuated between 27.5 °C and 30 °C during the year.

However, during most days of the year, the proposed optimal configuration of the PVPS met the water demand with a daily *LLP* of zero, except for 9 days. The maximum shortage occurred on December 24, when the values of *LLP*,solar radiation, water flow rate, and water deficit were 0.709, 1591.7 W/m^2, 8.732 m^3, and 21.269 m^3, respectively. By contrast, the minimum shortage occurred on November 9, when the values of *LLP*,solar radiation, water flow rate, and water deficit were 0.0074, 4197.2 W/m^2, 37.524 m^3, and 0.223 m^3, respectively.

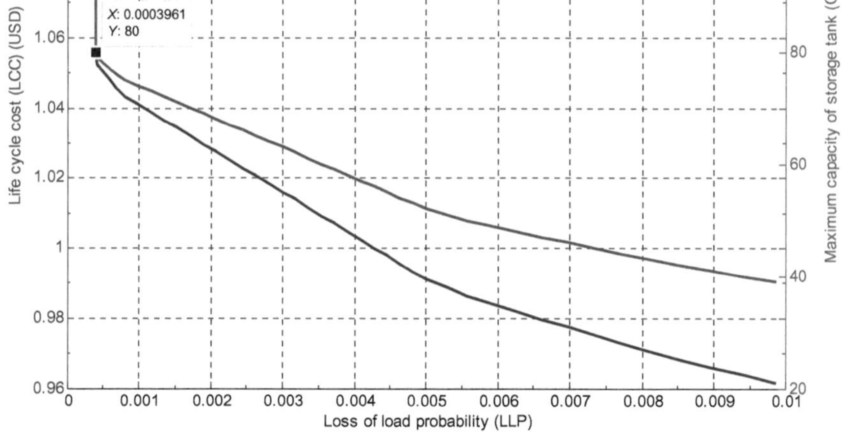

Fig. 5.24 Correlation between *LLP*, *LCC*, and C_n based numerical sizing method.

In spite of the daily productivity of system fulfilled the daily water demand on January 1, November 9, and December 22 and 25, but the daily *LLP* values of the system were not zero. The amount of resident water in the storage tank for the aforementioned dates was very little in the early hours of the morning. In particular, the amounts were 3.689 m^3 and 6.409 m^3 on November 9 and December 22, respectively, and the amount was zero on both January 1 and December 25.

The following MATLAB codes are used to execute the previous example

```
%%%%MATLAB Script of Diferential Evolution Multi-Objective(DEMO)algorithm
%%%% for optimizing the the size of PVPS (Ns, Np, Cn)
%%%% Last modification: 8/11/2015 - Sunday
clc;clear all;close all;
t1=cputime;
%%%%%%%%%%%%%%%%%//// Control parameter declaration ////%%%%%%%%%%%%%%
%%%%%
D=3;                                %Dimension of problem (5-parameters
                                     of PV module)
NP=50*D;                             %///&&&Size of pobulation (number of
                                      individuals)///&&&
%%%%%%%%%%%%%%%%%%%%%%%%%%%Mutation factor%%%%%%%%%%%%%%%%%%%%%%%%%%%
%%%%%
F=0.75;                                %///&&&Mutation factor///&&&
%%%%%%%%%%%%%%%%%%%%%%%%%%%%%%%%%Crossover rate%%%%%%%%%%%%%%%%%%%%%%%%
%%%%%
CR=0.5;                                 %///&&&Cross over rate factor///&&&
%%%%%%%%%%%%%%%%%%%%%Maximum number of generation (iteration)%%%%%%%%%%
%%%%%
GEN_max=100;                               %///&&&Maximum generation
                                            number///&&&
%%%%%%%% Reading the Solar radiation, Cell temperature, Module voltage
and Module current data %%%%%%%%
G=xlsread('Result_Sizing_DEMO_Method.xlsx',1,'I10:I4389');
%Reading the hourly solar radiation (W) (Year/10-4389),(Month/10-381),
   (Day/10-21)
Tc=xlsread('Result_Sizing_DEMO_Method.xlsx',1,'J10:J4389');
%Reading the hourly cell temperature (K)
Vm=xlsread('Result_Sizing_DEMO_Method.xlsx',1,'L10:L4389');
%Reading the hourly voltage of one module (V)
Im=xlsread('Result_Sizing_DEMO_Method.xlsx',1,'M10:M4389');
%Reading the hourly current of one module (A)
```

—cont'd

```
save ('input_variables_of_fitness_function', 'G','Tc','Im','Vm');
%Save G, Tc, Im & Vm data in MAT file
%%%%%%%%%%%%%%%%%%%%//// SAVING RESULTS IN MAT FILE ////%%%%%%%%%%%%%%
%%%%
file_name='Specifications_of DEMO.mat';
%Name of file to save control parameters of DEMO in MAT file
save (file_name,'NP','GEN_max','F','CR');
%save control parameters of DEMO in MAT file
%%%%%%%%%%%%%%%%%%%%//// SAVING RESULTS IN EXCEL FILE ////%%%%%%%%%%%%
%%%%%%
xlswrite ('Result_Sizing_DEMO_Method.xlsx',NP,1,'G4405');
xlswrite ('Result_Sizing_DEMO_Method.xlsx',GEN_max,1,'G4406');
xlswrite ('Result_Sizing_DEMO_Method.xlsx',F,1,'G4407');
xlswrite ('Result_Sizing_DEMO_Method.xlsx',CR,1,'G4408');
%%%//// DEFINING THE SEARCH SPACE OF DECISION VARIABLES (Ns, Np and Cn)
////%%%
Ns_l=4;              %Lower limit of no. of series PV modules (Ns)
Ns_h=70;              %Upper limit of no. of series PV modules (Ns)
Np_l=4;                %Lower limit of no. of parallel PV modules (Np)
Np_h=70;                %Upper limit of no. of parallel PV modules (Np)
Cn_l=35;                 %Lower limit of storage tank (Cn)
Cn_h=160;                  %Upper limit of storage tank (Cn)
L=[Ns_l Np_l Cn_l];          %Define lower limit vector of 3-decision-
                               parameters
H=[Ns h Np_h Cn_h];          %Define upper limit vector of 3-decision-
                               parameters
%%%%%%%%%%//// ALGORITHM'S VARIABLES AND MATRICES DECLARATION ////%%%%%
%%%%
x=zeros(D,1);                %Trial (offspring or candidate solution)
                              vector
pop=zeros(D,NP);              %Population matrix (parent solutions
                               matrix)
cand=zeros(D,NP);         %Population matrix (offspring or candidate or
                           trial solutions matrix)
fp1=zeros(1,NP);           %Overall first parent fitness function vector
                            of the population
fp2=zeros(1,NP);            %Overall second parent fitness function
                             vector of the population
fc1=zeros(1,NP);             %Overall first offspring fitness function
                              vector of the population
fc2=zeros(1,NP);     %Overall second offspring fitness function vector
                      of the population
r=zeros(3,1);          %Randomly selected indicies for mutation stage
```

Continued

—cont'd

```
%%%%%%%%%%%%%%%%%//// INITIOALIZATION THE POPULATION ////%%%%%%%%%%%%%
%%%%
for j=1:NP             %For all individuals vector
    for i=1:D           %For all variables of individual vector
        xx=L(i)+(H(i)-L(i))*rand(1,1);  %Initializing the individuals
                                         vector
        pop(i,j)=round(xx);              %Saving the individual
                                          vector in population
    end
    Ns=pop(1,j);                          %Specified the value of
                                           no. of series PV modules
                                           from population
    Np=pop(2,j);                %Specified the value of no. of parallel
                                 PV modules from population
    Cn=pop(3,j);                %Specified the value of size of storage
                                 tank from population
    [f1,f2]=fitness_function(Ns,Np,Cn); %Call the function of computing
                                         the fitness functions
    fp1(1,j)=f1;          %To save the overall first fitness function
                           for initial population
    fp2(1,j)=f2;          %To save the overall second fitness function
                           for initial population
end
popp=pop;             %To save the population of parents in popp
%%%%%%%%%%%%%%%%%%%%%%%%%%%//// Optimization ////%%%%%%%%%%%%%%%%%%%%%%%%%
for g=1:GEN_max                %For each generation (iteration)
    for j=1:NP                 %For each individual vector
        %%%%//// Selection three randomly indices for mutation ////%%%%
        %%%%%%%//// step to generate donor (mutation) vector ////%%%%%%
        r(1)=floor(rand*NP)+1;        %First random index
        while r(1)==j                  %To ensure ...
            r(1)=floor(rand*NP)+1;    %r(1)not equal j
        end
        r(2)=floor(rand*NP)+1;          %Second random index
        while (r(2)==j)||(r(2)==r(1))    %To ensure ...
            r(2)=floor(rand*NP)+1;        %r(2)not equal j and r(1)
        end
        r(3)=floor(rand*NP)+1;                          %Third random
                                                         index
        while (r(3)==j)||(r(3)==r(1))||(r(3)==r(2))     %To ensure ...
```

—cont'd

```
            r(3)=floor(rand*NP)+1;              %r(1)not equal j and r
                                                 (1)and r(2)
        end
        %%%%%%%%%%%%%%%%%%//// Mutation steps ////%%%%%%%%%%%%%%%%%%%%
        w=popp(:,r(3))+F.*(popp(:,r(1))-popp(:,r(2))); %To create the
mutation (donor) vector
        w=round(w);
        %%%%%%%%%%%%%%%%%%//// Crossover steps ////%%%%%%%%%%%%%%%%%%%%
        Rnd=floor(rand*D)+1;
        for i=1:D
            if (rand<CR)||(Rnd==i)
                x(i)=w(i);
            else
                x(i)=popp(i,j);
            end
        end
        % Checking the 3-decision-parameters of trial (offspring,
candidate) vector %
        %%%%%%%%%%%%//// with the boundary constraints ////%%%%%%%%%%%%
        for i=1:D
            if (x(i)<L(i))||(x(i)>H(i))
                x(i)=L(i)+(H(i)-L(i))*rand;
            end
        end
        Ns=x(1);                        %Specified the diode ideality factor
                                         value trial vector
        Np=x(2);                         %Specified the PV series resistance
                                          value trial vector
        Cn=x(3);                          %Specified the PV parallel
                                           resistance value trial vector
        [f1,f2]=fitness_function(Ns,Np,Cn);
        %Calculate the fitness functions for trial vector
        fc1(1,j)=f1;                           %To store the offspring's first
                                                fitness function
        fc2(1,j)=f2;                            %To store the offspring's
                                                 second fitness function
        cand(:,j)=x;                             %To store the offspring
                                                  solution in the cand matrix
 end                                             %End the loop for each
                                                   individual vectors
```

Continued

—cont'd

```
%%%Now we have two matrices of solutions, one of them for parent solutions
(popp) and the other for offspring or candidate (cand) solutions
%%%%%%%%%%%%%%%%%%%%%%%%%//// SELECTION STAGE ////%%%%%%%%%%%%%%%%%%%%%%%%%
%%%%%
%%%%%%%Comparison between parent and offspring solutions based on
pareto dominating concept%%%%%%%
l=1;                %Increment counter that used to indicate the index
                       to store some of offspring solutions beyond Np
pop=zeros(D,2*NP); %To create the population matrix to store pareto
                       dominating parents and offspring solutions
f1=zeros(1,2*NP);  %To create the first fitness function vector to
                       store first fitness of pareto dominating parents
                       and offspring solutions
f2=zeros(1,2*NP);  % To create the second fitness function vector to
                       store second fitness of pareto dominating parents
                       and offspring solutions
for i=1:NP
    if (((fp1(i)<=fc1(i))&&(fp2(i)<=fc2(i)))&&((fp1(i)<fc1(i))||(fp2(i)
                  <fc2(i)))) %If parent solution dominate offspring
                   (candidate) solution
        pop(:,i)=popp(:,i);     %store parent solution in the combined
                  population of size (D,Np to 2*Np)
        f1(1,i)=fp1(1,i);       %store first fitness function of parent
                  solution in the combined first fitness of size (1,Np
                  to 2*Np)
        f2(1,i)=fp2(1,i);       %store second fitness function of
                  parent solution in the combined second fitness of
                  size (1,Np to 2*Np)
    elseif (((fc1(i)<=fp1(i))&&(fc2(i)<=fp2(i)))&&((fc1(i)<fp1(i))||
                   (fc2(i)<fp2(i)))) %If offspring (candidate)
                   solution dominate parent solution
         pop(:,i)=cand(:,i);     %store offspring solution in the
                    combined population of size (D,Np to 2*Np)
         f1(1,i)=fc1(1,i);       %store first fitness function of
                   offspring solution in the combined first fitness
                   of size (1,Np to 2*Np)
         f2(1,i)=fc2(1,i);       %store second fitness function of
                   offspring solution in the combined second fitness
                   of size (1,Np to 2*Np)
  else                          %If parent and dominate offspring
                   (candidate) solutions are not dominated each other
```

—cont'd

```
        pop(:,i)=popp(:,i);     %store   parent   solution   in   the
                     combined population of size (D,Np to 2*Np)
        pop(:,NP+l)=cand(:,i);  %store offspring solution in the
                     combined population of size (D,Np to 2*Np) beyond
                     Np by l-value
        f1(1,i)=fp1(1,i);        %store  first  fitness  function  of
                     parent solution in the combined first fitness of
                     size (1,Np to 2*Np)
        f2(1,i)=fp2(1,i);        %store  second  fitness  function  of
                     parent  solution  in  the  combined  second  fitness
                     of size (1,Np to 2*Np)
        f1(1,NP+l)=fc1(1,i);    %store  first  fitness  function  of
                     offspring solution in the combined first fitness
                     of size (1,Np to 2*Np) beyond Np by l-value
        f2(1,NP+l)=fc2(1,i);    %store  second  fitness  function  of
                     offspring solution in the combined second fitness
                     of size (1,Np to 2*Np) beyond Np by l-value
        l=l+1;                  %Increment the value of l to prpare
                     the index for next offspring
    end
end
iDX=f2(1,:)==0;                   %To specify the index of redundant
                     locations in the second fitness function, pop and f1.
pop=pop(:,~iDX);                  %To truncate the redundant solutions
                     (with 0 value) from pop with index belong to iDX (in
                     the range (Np to 2*Np)
f1=f1(1,~iDX);                    %To truncate the redundant f1 of
                     solutions (with 0 value) from f1 with index belong
                     to iDX (in the range (Np to 2*Np)
f2=f2(1,~iDX);                    %To truncate the redundant f2 of
                     solutions (with 0 value) from f1 with index belong
                     to iDX (in the range (Np to 2*Np)
%%%%%% Nondominated Sorting, Note the no. of columns of f1, f2 and pop
                     is belong to the period [Np, 2Np] %%%%%%%%%%%%%%%
%%%%%%%%%%%%%% Sorting f1, f2 & pop Based Ascending Sorting of f1 %%%%%%%%%%%%%
[ff1, index]=sort(f1);                      %Sorting f1 in ascending
                                             sequence
m=size(f2,1);
ff2=f2(bsxfun(@plus,(index-1)*m,(1:m)'));   %Sorting  f2  based  on
                                             index
n=size(pop,1);
```

Continued

—cont'd

```
p=pop(bsxfun(@plus,(index-1)*n,(1:n)'));     %Sorting pop based on
                                              index
t=length(ff1);                               %To specify the length
                                              of sortting fitness
                                              function
%%%%%%% To swap the cases that has the same value for f1 %%%%%%%%
i=1;
while i<t
    for j=i:t-1
        if (ff1(1,i)==ff1(1,j)) && (ff2(1,j)<ff2(1,i))
           a=ff2(1,j);
           ff2(1,j)=ff2(1,i);
           ff2(1,i)=a;
           b=p(:,j);
           p(:,j)=p(:,i);
           p(:,i)=b;
        end
    end
    i=i+1;
end
%%%%%%%%%% Nondominated Ranking %%%%%%%%%%%%%%%%%%%%%%%%%%%%%%%%%%%
%%%%%%%%%%%%%
rk=1;                   %No. of rank
e=0;                    %Counter for nondominated ranking sorted_f1,
                         sorted_f2 and set
ee=0;                    %Counter for counting the number of solutions
                          have the same rank
Q=length(ff1);           %To specify the length of ff1
sorted_f1=zeros(1,Q);     %To create the sorted_f1 vector to store the
                           nondominated sorting f1
sorted_f2=zeros(1,Q);      %To create the sorted_f2 vector to store the
                            nondominated sorting f2
set=zeros(D,Q);         %To create the set matrix to store the
                         nondominated sorting population
I=zeros(1,Q);
%To create vector I to specify the index of solution that belong to the
   rank level rk
N=zeros(1,Q);
%To create vector N to store the number of solutions those belong to
   the rank level rk
```

—cont'd

```
while Q~=0
    Sc_fitness1=ff1(1,1);
    Sc_fitness2=ff2(1,1);
    Rank_Sc=rk;
    e=e+1;
    ee=ee+1;
    sorted_f1(1,e)=Sc_fitness1;
    ff1(1)=[];                  %To truncate the first solution in ff1
    sorted_f2(1,e)=Sc_fitness2;
    ff2(1)=[];                  %To truncate the second solution in ff2
    set(:,e)=p(:,1);            %To copy the first solution to set
                                 matrix
    p(:,1)=[];                  %To truncate the first solution in p
    Population=p(:,:);
    %Save p matrix in population matrix to deal with it for the
     remainder steps
    Fitness1=ff1;
    %To save ff1 in Fitness1 to deal with it for the remainder steps
    Fitness2=ff2;
    %To save ff2 in Fitness2 to deal with it for the remainder steps
      for i=1:Q-1
          Sn_fitness1=Fitness1(1,i);
          Sn_fitness2=Fitness2(1,i);
          if (Sc_fitness1==Sn_fitness1) && (Sc_fitness2==Sn_fitness2)
             Rank_Sn=rk;
             e=e+1;
             ee=ee+1;
             sorted_f1(1,e)=Sn_fitness1;
             sorted_f2(1,e)=Sn_fitness2;
             set(:,e)=Population(:,i);
             I(1,e)=i;                %To store the index of solution
          elseif Sn_fitness2<Sc_fitness2
             Rank_Sn=rk;
             e=e+1;
             ee=ee+1;
             sorted_f1(1,e)=Sn_fitness1;
             sorted_f2(1,e)=Sn_fitness2;
             set(:,e)=Population(:,i);
             I(1,e)=i;                %To store the index of solution
          end
          Sc_fitness1=Sn_fitness1;
          Sc_fitness2=Sn_fitness2;
      end
```

Continued

—cont'd

```
  Q=Q-ee;          %To specify the residual solutions that not ranked
  N(1,rk)=ee;      %To store the number of solutions in the rank level rk
  rk=rk+1;         %To increment the rank level
  ee=0;            %To reset the number of solutions counter for the
                    next rank level
  INDEX=I(1,:)==0;  %To specify the index of zero elements in I
  I=I(1,~INDEX);     %To truncate the elements of I that have index
                      values in INDEX vector
  p(:,I)=[];          %To truncate the solutions those ranked in rk
                       level
  ff1(I)=[];     %To truncate the fitness functions of solutions those
                  ranked in rk level
  ff2(I)=[];     %To truncate the fitness functions of solutions those
                  ranked in rk level
  I=zeros(1,Q); %To resize I based on the residual solutions those not
                 sorted
end
idx=N(1,:)==0; %To specify the index of zero elements in N
N=N(1,~idx);   %To truncate the elements of I that have index values
                in idx vector
k=length(N);   %To specify the total no. of ranking levels
%%%%%%Crowding Distance Ranking %%%%%%%%%%%%%
%%%%Part 1 %%%%
f1max=max(sorted_f1);            %To specify the maximum value for f1
f2max=max(sorted_f2);            %To specify the maximum value for f2
f1min=min(sorted_f1);            %To specify the minimum value for f1
f2min=min(sorted_f2);            %To specify the minimum value for f2
SN=zeros(1,length(N));           %To create SN vector that to
 specify which front level set of solutions) should be ranked based
 on crowding distance
%%%%%%%%To specify which front level should be ranked based on
crowding distance%%%%%%%
sum=0;
for i=1:k
    sum=sum+N(i);
    if sum>NP
        SN(i)=1;
        break
    end
end
```

—cont'd

```
%%%%% Part 2, To specify the c value that used to do crowding distance
%%%%%%%%
popn=zeros(D,NP);
f1n=zeros(1,NP);
f2n=zeros(1,NP);
c=1;
sum=0;
for i=1:k
    if SN(i)==0
        NN=N(i);
        kk=sum+1;
        sum=sum+NN;
        popn(:,kk:sum)=set(:,kk:sum);
        %To save the set of solutions that should already be in the new
           population for next generation
        f1n(:,kk:sum)=sorted_f1(:,kk:sum); %To save the fitness function
           of set of solutions that should already be in the new population for
           next generation
        f2n(:,kk:sum)=sorted_f2(:,kk:sum); %To save the fitness function
           of set of solutions that should already be in the new population for
           next generation
        c=c+1;                                %Increment c
        if sum==NP
            c=0;                              %If size of sets of
        solutions should already be in the new population equal to Np
        then reset c
        end
    else
       break  %If SN(i)==1 then stop the loop
    end
end
%%%%%%% Part 3, To do crowding distance for the set that specified by c
%%%%%%%%%
if c==1    %If the size of first set that belong to the first front
             level is greater than Np
    NN=N(c);
    if N(c)>NP
        sett=set(:,1:NN);
        f1s=sorted_f1(1,1:NN);
        f2s=sorted_f2(1,1:NN);
```

Continued

—cont'd

```
        [settn, f1ss, f2ss, CD]=Crowding_Distance_Function(NN, sett,
f1s, f2s, f1max, f1min, f2max, f2min);
        popn(:,1:NP)=settn(:,1:NP);
        %To store the required solutions that ranked based on the
        crowding distance
        f1n(1,1:NP)=f1ss(1,1:NP);
        %To store the fitness function values of required solutions
        that ranked based on the crowding distance
        f2n(1,1:NP)=f2ss(1,1:NP);
        %To store the fitness function values of required solutions
        that ranked based on the crowding distance
    else
        %If the size of first set that belong to the first front level
        is equal to Np
        popn(:,:)=set(:,1:NN);
        f1n(1,1:NP)=sorted_f1(1,1:NN);
        f2n(1,1:NP)=sorted_f2(1,1:NN);
    end
 elseif c>1            %If the size of first set that belong to the
        first front level is less than Np, so we add other solutions
        belong to other front levels
    sum=0;
    for i=1:(c-1)
        sum=sum+N(i);
    end
    NN=N(c);
    if (sum+NN)>NP %If the size of (sum+NN) sets of solutions those
        belong to various front levels is greater than Np
        sett=set(:,(sum+1):(sum+NN));
        f1s=sorted_f1(1,(sum+1):(sum+NN));
        f2s=sorted_f2(1,(sum+1):(sum+NN));
        [settn, f1ss, f2ss, CD]=Crowding_Distance_Function(NN, sett,
 f1s, f2s, f1max, f1min, f2max, f2min); %Crowding distance function
        popn(:,(sum+1):NP)=settn(:,1:NP-sum); %To store the required
         solutions that ranked based on the crowding distance
        f1n(1,(sum+1):NP)=f1ss(1,1:NP-sum); %To store the fitness
         function values of required solutions that ranked based on
         the crowding distance
        f2n(1,(sum+1):NP)=f2ss(1,1:NP-sum); %To store the fitness
         function values of required solutions that ranked based on
         the crowding distance
    else                                      %If the size of (sum+NN)
        sets of solutions those belong to various front levels is equal to Np
```

—cont'd

```
        popn(:,sum+1:NP)=set(:,(sum+1):(sum+NN));
        f1n(1,sum+1:NP)=sorted_f1(1,(sum+1):(sum+NN));
        f2n(1,sum+1:NP)=sorted_f2(1,(sum+1):(sum+NN));
    end
else
    popn;
    f1n;
    f2n;
end
pop=popn; %To store set of solutions with size Np in pop for next
          generation purpose
f1=f1n;    %To store the fitness function values in f1 for next
           generation purpose
f2=f2n;     %To store the fitness function values in f2 for next
            generation purpose
%%%%% PREPARING THE NEW POPULATION AND ITS FITNESS FUNCTIONS (f1 & f2)
for next generation
popp=pop;
%To represent the new population as population of parents solutions
 for next generation
fp1=f1; %To represent the new f1 as f1 of parents solutions for next
        generation
fp2=f2; %To represent the new f2 as f2 of parents solutions for next
        generation
end      %End the loop for each generation (iteration)
%%%%%%//// Results of optimization of 3-parameters of PV module ////%%%
%%%%
Ns_vector=popp(1,:)';
Np_vector=popp(2,:)';
Cn_vector=popp(3,:)';
N_vector=Ns_vector.*Np_vector;
LLP=fp1';
LCC=fp2';
LP=20;
%Life time of PVPS
Cost_per_year=LCC./LP;
%To compute the cost of PVPS per one year
save(file_name,'N_vector','Ns_vector','Np_vector','Cn_vector','fp1','
fp2','Cost_per_year'); %To save the fitness function in MAT file for
rr run
```

Continued

—cont'd

```
xlswrite ('Result_Sizing_DEMO_Method.xlsx',N_vector,2,'AA2');
%To save the vector of N values in excel sheet
xlswrite ('Result_Sizing_DEMO_Method.xlsx',Ns_vector,2,'AB2');
%To save the vector of Ns values in excel sheet
xlswrite ('Result_Sizing_DEMO_Method.xlsx',Np_vector,2,'AC2');
%To save the vector of Np values in excel sheet
xlswrite ('Result_Sizing_DEMO_Method.xlsx',Cn_vector,2,'AD2');
%To save the vector of Cn values in excel sheet
xlswrite ('Result_Sizing_DEMO_Method.xlsx',LLP,2,'AE2');
%To save the vector of f1 (LLP) values in excel sheet
xlswrite ('Result_Sizing_DEMO_Method.xlsx',LCC,2,'AF2');
%To save the vector of f2 (LCC) values in excel sheet
xlswrite ('Result_Sizing_DEMO_Method.xlsx',Cost_per_year,2,'AG2');
%To save the vector of cost per year (LCC/year) values in excel sheet
e=cputime-t1
```

Matlab code for crowding distance function

```
%%%% MATLAB Script function for crowding distance of
%%%% Diferential Evolution Multi-Objective (DEMO)algorithm
%%%% for optimizing the 3-decision-parameters (Ns, Np and Cn)
%%%% Lat modification: 9/11/2015 - Mondday
function [settn, f1ss, f2ss, CD]=Crowding_Distance_Function(NN, sett,
f1s, f2s, f1max, f1min, f2max, f2min)
[f11s, indices1]=sort(f1s);
[f22s, indices2]=sort(f2s);
ff2=f2s(indices1);
m=size(sett,1);
sett_sorted=sett(bsxfun(@plus,(indices1-1)*m,(1:m)'));
d1=zeros(1,NN);
d2=zeros(1,NN);
d=zeros(1,NN);
d1(1)=inf;
d2(1)=inf;
d1(NN)=inf;
d2(NN)=inf;
for i=2:NN-1
    d1(i)=(f11s(i+1)-f11s(i-1))/(f1max-f1min);
    d2(i)=(f22s(i+1)-f22s(i-1))/(f2max-f2min);
end
```

—cont'd

```
d(1,1)=inf;
d(1,NN)=inf;
for j=1:NN
    for i=1:NN
        if indices1(i)==indices2(j)
            d(j)=d1(j)+d2(i);
        end
    end
end
[CD, index]=sort(d, 'descend');
n=size(sett_sorted,1);
settn=sett_sorted(bsxfun(@plus, (index-1)*n, (1:n)'));
f1ss=f11s(index);
f2ss=ff2(index);
```

Matlab code for the utilized fitness function

```
%%%% MATLAB Script function for computing the fitness functions of
%%%% Diferential Evolution Multi-Objective (DEMO)algorithm
%%%% for optimizing the 3-decision-parameters (Ns, Np and Cn)
%%%% Lat modification: 9/11/2015 - Mondday
function [f1,f2]=fitness_function(Ns,Np,Cn)      %%Define a function to
compute the fitness functions
%%%%%%%%%%%%%%%%%//// Declaration the constants ////%%%%%%%%%%%%%%%%%
%%%%%
load  ('input_variables_of_fitness_function',  'G','Tc','Im','Vm'); %
Loading G, Tc, Im & Vm data from MAT file
%%%%%%%%%%%%%//// COMPUTING FIRST FITNESS FUNCTION (LLP) ////%%%%%%%%%
%%%%%%%%%%%%
%%%%%%%%%%%%%%%%% h1 and h2 are changed according to pumping head %%%%%
%%%%%%%%%%%%
%%h1=20, 30, 40, 50 and 60, respectively.
%%h2=0.1444, 0.1907, 0.2371, 0.2835 and 0.3298, respectively
h1=20;                       %%%&&&Factors of head equation we have got
                              it from head calculations
h2=0.1444;                    %%%&&&Factors of head equation we have
                               got it from head calculations
```

Continued

—cont'd

```
%%%%%%%%%%%%%%%%%%%%%%%%%%%%% PV ARRAY %%%%%%%%%%%%%%%%%%%%%%%%%%%%
%%%%%
Va=Ns.*Vm;                  %Computing the hourly voltage of PV array (V)
Ia=Np.*Im;                  %Computing the hourly current of PV array (A)
Vpv=Va;                     %Output voltage of PV array for storing purpose
Ipv=Ia;                     %Output current of PV array for storing purpose
Pao=Va.*Ia;                 %Computing the hourly output power of PV array (W)
Am=0.9291;                  %The area of one PV module (m^2)
A=Ns*Np*Am;                 %Computing the area of PV array (m^2)
Pai=A.*G;                   %Computing the hourly input power of PV
                             array (W)
Pai1=(Pai==0).*1;
Pai2=Pai+Pai1;
effa=Pao./Pai2;             %Computing the hourly efficiency of PV
array
%%%%%%%%%%%%%%%%%%%%%%%%%%%%% MOTOR %%%%%%%%%%%%%%%%%%%%%%%%%%%
%%%%%
effconv=0.9;                %DC-DC converter efficiency
Va=0.95.*Va;                %The output voltage of DC-DC converter
Ia=0.9.*Ia;                 %The output current of DC-DC converter
Ra=0.8;                     %Armature resistance of DC motor (Ohm)
Km=0.175;                   %Torque and back emf constant
                             (V/(rad/sec))
Ebb=Va-(Ra.*Ia);            %Computing the hourly back emf
                             voltage of motor (V)
%%%%%%%%% The case of overcurrent supplied to motor by PV array %%%%%%%
Eb=(Ebb>=0).*Ebb;     %Set Eb=0 when Ebb<0 (overcurrent)/turn off motor
Ia=(Ebb>=0).*Ia;      %Set Ia=0 when Ebb<0 (overcurrent)/turn off motor
Va=(Ebb>=0).*Va;      %Set Va=0 when Ebb<0 (overcurrent)/turn off motor
%%%%%%%%%%%%%%%%%%%%%%%%%%%%%%%%%%%%%%%%%%%%%%%%%%%%%%%%%%%%%%%%%%
%%%%
Tm=Km.*Ia;                  %Computing the hourly torque of DC motor
Tmm=(Tm==0).*1;
Tm1=Tmm+Tm;
Rou=1000;                   %Density of water (Kg/m^3)
g=9.81;                     %Acceleration due to gravity (m/Sec^2)
d1=33.5*0.001;              %Inlet impeller diameter (mm)
d2=160*0.001;               %outlet impeller diameter (mm)
beta1=38*2*pi/360;          %Inclination angle of impeller blade at
                             impeller inlet (degree)
```

—cont'd

```
beta2=33*2*pi/360;                 %Inclination angle of impeller blade
                                    at impeller outlet (degree)
b1=5.4*0.001;                      %Height of impeller blade at impeller
                                    inlet (mm)
b2=2.2*0.001;                      %Height of impeller blade at impeller
                                    outlet (mm)
Kp=Rou*2*pi*b1*(d1/2)^2*tan(beta1)*((d2/2)^2-((b1*(d1/2)^2*tan
(beta1))/(b2*tan(beta2))));
%Computing the hourly output power of DC motor (W)
Pdev=Eb.*Ia;                       %Computing the hourly development
                                    power by motor (W)
Omega=abs(sqrt((Km.*Ia)./Kp));     %Computing the hourly angular speed
                                    of motor (rad/sec)
Pmo=Pdev;                          %By neglecting rotational losses,
Pmi=Pao.*0.9;                      %Computing the hourly input power
                                    of DC motor (W)
PMI1=(Pmi==0).*1;                  %To overcome divided by zero
PMI2=Pmi+PMI1;                     %To overcome divided by zero
effm=Pmo./PMI2;                    %Computing the hourly efficiency
                                    of DC motor
%%%%%%%%%%%%%%%%%%%%%%%%%%%%%% PUMP %%%%%%%%%%%%%%%%%%%%%%%%%%%%%%%%%%%
Tp=Tm;                  %The produced torque by motor is equal the torque
                         required for pump (Nm)
Eh=Tp.*Omega;           %Computing the hydraulic energy (W)
Ppo=Eh;                 %Computing the hourly output power of pump (W)
Ppo=(Ebb>=0).*Ppo;
Ppi=Pmo;                %Computing the hourly input power of pump (W)
PPI1=(Ppi==0).*1;       %To overcome divided by zero
PPI2=Ppi+PPI1;          %To overcome divided by zero
effpp=Ppo./PPI2;        %Computing the hourly efficiency of pump
effp=effpp;
Q=zeros(length(Eh),1);
for ii=1:length(Eh)
    r1=h2*2.725;                    %Computing the flow rate of water
    r2=0;                           %Computing the flow rate of water
    r3=h1*2.725;                    %Computing the flow rate of water
    r4=-Eh(ii);                     %Computing the flow rate of water
    r=roots([r1 r2 r3 r4]);         %Computing the flow rate of water
    if (imag(r(1))==0 && real(r(1))>0)  %Choosing the real value of
                                         the flow rate of water
        QQQ=real(r(1));
```

Continued

—cont'd

```
    elseif (imag(r(2))==0 && real(r(2))>0)    %Choosing the real value of
                                               the flow rate of water
    QQQ=real(r(2));
    elseif (imag(r(3))==0 && real(r(3))>0)    %Choosing the real value of
                                               the flow rate of water
    QQQ=real(r(3));
    else
    QQQ=0; %If all the roots are complex and/or the real part is negative
           number or zero
    end
    QQ(ii,1)=QQQ; %Hourly flow rate (m^3/h)
end
Q1=(QQ==0).*1;    %To overcome divided by zero
Q2=QQ+Q1;         %To overcome divided by zero
H=Eh./(2.725.*Q2); %Computing the head of pumping water (m)
%%%%%%%%%% OVERALL SYSTEM %%%%%%%%%%%%%%%%%%%%%%%%%%%%%%%%%%%%%%%%%%%
%%%%
effsub=effm.*effp.*effconv;    %Computing hourly subsystem efficiency
effoverall=effa.*effm.*effp.*effconv; %Computing hourly overall
                                       efficiency
QQ=(Ebb>=0).*QQ;
QQ=(effpp<=0.95).*QQ;
Q=[0;QQ];                              %To add initial case Q=0 for
                                        programming purposes
d=2.5;                                  %Hourly demand water (m^3/h)
Cr=zeros(length(Q),1);           %To specify the size of current
                                  resident matrix of storage tank
Qexcess_pv=zeros(length(Q),1);  %To specify the size of excess water
matrix
Qexcess_s=zeros(length(Q),1);
SOC=zeros(length(Q),1);
Qdef_pv=zeros(length(Q),1);       %To specify the size of deficit
                                   water matrix (before tank)
Qdeficit_s=zeros(length(Q),1);     %To specify the size of deficit
                                    water matrix (after tank)
X=Q(2:end,1)-d;                      %Difference between the hourly
                                      production and demand water
%%%%%%%%%%%%%********** Before Tank **********%%%%%%%%%%%%%%%
Qdef_pv(2:end,1)=(X<0).*abs(X);        %Computing hourly deficit water
                                        before tank (m^3)
```

—cont'd

```
%%%%%%%%%%%%%********** After Tank **********%%%%%%%%%%%%%
C=length(Q)-1;
Qexcess_pv(2:end,1)=(X>=0).*abs(X);  %Computing hourly excess water
                                      before and after tank (m^3)
for i=1:C
    Cr(i+1,1)=((Cr(i,1)+X(i,1))>=0).*abs(Cr(i,1)+X(i,1));
%To compute the hourly current resident water in the tank (m^3)
    SOC(i+1,1)=Cr(i+1,1)/Cn;  %Computing the hourly state of charge of
                               storage tank
    if SOC(i+1,1)>=1
        SOC(i+1,1)=1;
        Qexcess_s(i+1,1)=Cr(i+1,1)-Cn;
        Cr(i+1,1)=Cn;
    else
        Qexcess_s(i+1,1)=0;
    end
    Qdeficit_s(i+1,1)=((Cr(i,1)+X(i,1))<0).*abs(Cr(i,1)+X(i,1));
%To compute the hourly deficit water (m^3/h) (after tank)
end
Q=Q(2:end,1);                      %Final computing of hourly flow rate
                                    of water (m^3)
Qexcess_pv=Qexcess_pv(2:end,1);    %Final computing of hourly excess
                                    water before /after tank (m^3)
Qexcess_s=Qexcess_s(2:end,1);      %Final computing of hourly excess
                                    water after the tank is filled (m^3)
%%f3=Qexcess_s;
Qdeficit_s=Qdeficit_s(2:end,1);    Final computing of hourly deficit
                                    water after tank (m^3)
Qdef_pv=Qdef_pv(2:end,1);          %Final computing of hourly deficit
                                    water before tank (m^3)
Cres=Cr(2:end,1);                  %Final computing of hourly current
                                    resident water in tank (m^3)
SOC=SOC(2:end,1);                  %Final computing of hourly state of
                                    charge (SOC)
D=zeros(length(Q),1)+d;            %Constructing the matrix of hourly
                                    demand water (m^3)
LLP=sum(Qdeficit_s(1:end,1))/sum(D(1:end,1)); %Computing the LLP of one
                                                year
f1=LLP;                                       %First fitness function
```

Continued

—cont'd

```
%%%%%%%%%//// COMPUTING SECOND FITNESS FUNCTION (f2=LCC) ////%%%%%%%
%%%%%
CAc=800;               %Total capacity of converter required for system (W)
CAmp=840;              %Total capacity of motor-pump set required for
system (W)
UCpv=1;                %%Unit cost of PV ($/Wp)
UCc=0.5;               %%Unit cost of converter ($/W)
UCmp=0.75;             %%Unit cost of motor-pump set ($/W)
UCt=20;                %%Unit cost of storage tank ($/m^3)
ICI=4000;              %%Civil and installation works cost ($)
FR=0.04;               %%Inflation rate
IR=0.08;               %%Interest rate
LP=20;                 %%Life time of PVPS
Nr=1;    %%Number of replacement times for motor-pump set and converter
N=Ns*Np;               %Total number of PV modules
%%%%%%% COST SCRIPT %%%%%%%%%%%%%%%%%%
CApv=120*N;            %Total capacity of PV required for system (Wp)
CAt=Cn;                %Total capacity of storage tank required for
                        system (m^3)
%%%%%%%%%%%%%%%%% COMPUTING INITIAL COST OF PVPS %%%%%%%%%%%%%%%%%%%%
IC=(CApv*UCpv)+(CAc*UCc)+(CAmp*UCmp)+(CAt*UCt)+ICI; %Initial cost of
PVPS ($)
%%%%%%%%%%%%%%%%% COMPUTING REPLACEMENT COST OF PVPS %%%%%%%%%%%%%%%%
RCCmp=zeros(1,2);     %Array to acc the replacement cost for every
                       replacement of motor-pump set
RCCc=zeros(1,2);    %Array to accumulate the replacement cost for every
                     replacement of converter
for j=1:Nr
    RRCmp=(CAmp*UCmp)*(((1+FR)/(1+IR))^((LP*j)/(Nr+1))); %Replacement
       cost of motor-pump set ($)
    RRCc=(CAc*UCc)*(((1+FR)/(1+IR))^((LP*j)/(Nr+1))); %Replacement
       cost of motor-pump set ($)
    RCCmp(1,j)=RRCmp;          %Accumulate the replacement costs for
                                motor-pump set
    RCCc(1,j)=RRCc;            %Accumulate the replacement costs for
                                converter
end
RCmp=sum(RCCmp);      %Sum the replacement cost for L replacement motor-
                        pump set ($)
RCc=sum(RCCc);        %Sum the replacement cost for L replacement
                        converter ($)
RC=RCmp+RCc;          %Computing the replacement cost for PVPS ($)
```

—cont'd

```
%%%%%%%%%%%%%%%% COMPUTING OPERATION AND MAINTENANCE COST OF PVPS %%%%
%%%%
MCpv0=0.01*(CApv*UCpv);  %Maintenance and operation cost of PV in the
                          first year ($)
MCmp0=0.03*(CAmp*UCmp);  %Maintenance and operation cost of motor-pump
                          set in the first year ($)
MCt0=0.01*(CAt*UCt);     %Maintenance and operation cost of storage
                          tank in the first year ($)
MCpv=MCpv0*((1+FR)/(IR-FR))*(1-((1+FR)/(1+IR))^LP);
%Maintenance and operation cost of PV along life time of PVPS ($)
MCmp=MCmp0*((1+FR)/(IR-FR))*(1-((1+FR)/(1+IR))^LP);
%Maintenance and operation cost of motor-pump set along life time of
PVPS ($)
MCt=MCt0*((1+FR)/(IR-FR))*(1-((1+FR)/(1+IR))^LP);
%Maintenance and operation cost of storage tank along life time of PVPS ($)
MC=MCpv+MCmp+MCt;
%Calculating the maintenance and operation costs of PVPS along life
time of PVPS ($)
%%%%%%%%%%%%%%%% COMPUTING LIFE CYCL COST OF PVPS %%%%%%%%%%%%%%%%%%%%%
LCC=IC+RC+MC;             %Computing the LCC of PVPS ($)
f2=LCC;                   %Second fitness function
end
```

5.4 Multicriteria decision-making method-based sizing of PVPS

In this section, the sizing of the PVPS is represented as a multicriteria decision-making (MCDM) problem. In any MCDM problem, there are three fundamental terms that should be defined: alternatives, criteria, and decision matrix (*DM*). The alternatives are represented by the set of configurations of the PVPS. The configuration of the PVPS comprises the number of PV modules connected in series (N_s), the number of PV modules connected in parallel (N_s), and the capacity of the storage tank (C_n). The last term that should exist to solve an MCDM problem is the *DM*, which combines *m* alternatives and *n* criteria. The elements of *DM* are the performance of the alternatives in terms of each criterion.

The random nature of solar energy is one of the main obstacles in designing an effective PVPS. Thus, an optimal and effective sizing approach is essential to ensure satisfactory performance. In this section, a technique for order performance by the similarity to ideal solution (TOPSIS) method integrated with the analytic hierarchy process (AHP) method is proposed to optimally size a PVPS based on technoeconomic aspects. The loss of load probability (LLP) and excess water volume are considered technical criteria, whereas the life cycle cost (LCC) is represented as an economic criterion to size the system. An iterative program is used to compute the performance of each alternative (PVPS configuration) in terms of each criterion. After that, the hybrid AHP-TOPSIS sorts the PVPS configurations from the best to worst based on predefined weights for each criterion. The optimal configurations are sorted in terms of minimization of the *LLP*, *LCC*, and Q_e simultaneously. For brevity, the results of the numerical sizing example presented in Chapter 4 are used in the current example. Therefore, only the hybrid AHP-TOPSIS methods are discussed in the following sections.

5.4.1 AHP-TOPSIS model for sizing PVPS

A hybrid model is used in this example to sort the configurations of the PVPSs from the best to the worst ones. The model combines AHP and TOPSIS methods. The AHP method is used to nominate appropriate weights for each criteria according to Saaty's scale. After that, the weights are used with the TOPSIS method to make a decision about the optimal configuration. Fig. 5.23 shows the hybrid AHP-TOPSIS model that is proposed to select the optimal configuration of the PVPS. The following sections illustrate how the weights are derived by AHP, and TOPSIS will be discussed in detail.

5.4.1.1 AHP method for developing weights

The weights are essential in most MCDM methods to show the importance of each criteria that dominate the MCDM problem. AHP is one of the most widely used MCDM methods. AHP is used to derive a preference rating among the criteria that dominate the MCDM problem by using pairwise comparisons. The total number of required pairwise comparisons for a problem that has n evaluation criteria are $n * (n-1)/2$. As was previously stated, the criteria used to solve the sizing problem of the PVPS as an MCDM problem are *LLP*, *LCC*, and Q_e. The nine-point scale method is used to compare the two criteria. The comparison between the criteria is to illustrate how many times one criteria is more important than another. A survey based on the opinion of three evaluators (experts) is achieved to complete the pair-wise comparisons between the criteria. The preferences and evaluation of the evaluators are illustrated in Table 5.3. The evaluations of the evaluator were as follows. The first evaluator strongly favors *LLP* over *LCC*, slightly favors *LLP* over Q_e, and strongly favors Q_e over *LCC*. For the second evaluator, slight favor is given for *LLP* over *LCC*, strong favor for *LLP* over Q_e, and very strong favor for Q_e over *LCC*. The third evaluator gave very strong favor for *LLP* over *LCC* and Q_e,respectively, and no favor between *LCC* and Q_e.

A comparison matrix is constructed, as shown in Table 5.4, based on the evaluator's preferences. After that, each element of each column is divided by the sum of the element values of the column to constitute the elements of the normalized matrix. The element of the aggregation vector is equal to the sum of the element weight values of each criterion by normalizing the corresponding element values of the aggregation vector. It is worth mentioning that the total sum of weights for all criteria should be governed by

$$\sum_{j=1}^{n} w_j = 1, \tag{5.19}$$

where w_j is the weight of the jth criteria and n is the number of criteria that dominate the MCDM problem. Table 5.4 represents the steps of the AHP method to measure the weights based on the preferences of the three evaluators.

5.4.1.2 Technique for order preference by similarity to ideal solution (TOPSIS)

In TOPSIS, alternatives are sorted according to the shortest distance from the ideal solution and the farthest distance from the negative ideal solution in some geometrical sense. The TOPSIS method can be described by the following steps.

An MCDM problem can be easily expressed and solved in the matrix format. Therefore, a decision matrix (*DM*) that consists of m alternatives and n criteria should be created. The elements of *DM* (a_{ij}) represent the performance of alternative A_i when it is evaluated in terms of criterion C_j, where $i=1, 2, \ldots, m$ and $j=1, 2, \ldots, n$. It is also assumed that the weights of criterion C_j are obtained from the AHP method and are denoted by w_j. Thus, the *DM*of an MCDM problem is given by,

Table 5.3 Pairwise comparisons of three evaluators based on Saaty's scale.

		Extremely favor	Very strong favor	Strong favor	Slightly favor	Equal	Slightly favor	Strong favor	Very strong favor	Extremely favor	
Evaluator	Criteria	9	7	5	3	1	3	5	7	9	Criteria
1	*LLP*			✓							*LCC*
	LLP				✓						Q_e
	LCC							✓			Q_e
2	*LLP*				✓						*LCC*
	LLP			✓							Q_e
	LCC								✓		Q_e
3	*LLP*		✓								*LCC*
	LLP		✓								Q_e
	LCC					✓					Q_e

Table 5.4 AHP processing matrix to compute the weights of criteria.

		Original matrix			Normalized matrix				
Evaluator	Criteria	*LLP*	*LCC*	Q_e	*LLP*	*LCC*	Q_e	Aggregation	Weight
1	*LLP*	1.00	5.00	3.00	0.65	0.45	0.71	1.82	0.61
	LCC	0.20	1.00	0.20	0.13	0.09	0.05	0.27	0.09
	Q_e	0.33	5.00	1.00	0.22	0.45	0.24	0.91	0.30
	Sum	1.53	11.00	4.20				3.00	
2	*LLP*	1.00	3.00	5.00	0.65	0.27	0.81	1.74	0.58
	LCC	0.33	1.00	0.14	0.22	0.09	0.02	0.33	0.11
	Q_e	0.20	7.14	1.00	0.13	0.64	0.16	0.93	0.31
	Sum	1.53	11.14	6.14				3.00	
3	*LLP*	1.00	7.00	7.00	0.78	0.78	0.78	2.33	0.78
	LCC	0.14	1.00	1.00	0.11	0.11	0.11	0.33	0.11
	Q_e	0.14	1.00	1.00	0.11	0.11	0.11	0.33	0.11
	Sum	1.29	9.00	9.00				3.00	

$$DM = \begin{array}{c} \\ \\ A_1 \\ \vdots \\ A_m \end{array} \begin{array}{c} C_1 \;\cdots\cdots\cdots\; C_n \\ w_1 \;\cdots\cdots\cdots\; w_n \\ \begin{bmatrix} a_{11} & \cdots\cdots\cdots & a_{1n} \\ \vdots & \cdots\cdots\cdots & \vdots \\ a_{m1} & \cdots\cdots\cdots & a_{nm} \end{bmatrix} \end{array}, \tag{5.20}$$

Step 1: Construction of a normalized decision matrix

In this step, the various criteria with dimensions (various units) are converted into nondimensional criteria. In other words, the decision matrix DM is normalized to obtain the normalized decision matrix R, where each element in DM is normalized by

$$r_{ij} = \frac{a_{ij}}{\sqrt{\sum_{k=1}^{m} \left(a_{ij}\right)^2}}, \tag{5.21}$$

As a result of step 1, the R matrix will be obtained as described below

$$R = \begin{bmatrix} r_{11} & \cdots & r_{1n} \\ \vdots & \ddots & \vdots \\ r_{m1} & \cdots & r_{mn} \end{bmatrix}, \tag{5.22}$$

Step 2: Construction of a weight normalized decision matrix

The weights vector obtained by the AHP method is used with the R matrix to constitute a weighted normalized decision matrix that is denoted by V. The columns of the V matrix are calculated by multiplying each column of matrix R with its associated weight w_j from the weights vector as given by.

$$v_{ij} = w_j r_{ij}, \text{ for } i = 1, 2, \ldots, m \text{ and } j = 1, 2, \ldots, n \tag{5.23}$$

The outcome of the current step is matrix V, which is explained below

$$V = \begin{bmatrix} v_{11} & \cdots & v_{1n} \\ \vdots & \ddots & \vdots \\ v_{m1} & \cdots & v_{mn} \end{bmatrix}, \tag{5.24}$$

Step 3: Computing the ideal and the negative ideal solutions

In this step, two fictitious solutions A^*(the ideal solution) and A^- (the negative ideal solution) are defined as;

$$\begin{aligned} A^* &= \left\{ \left(\max_i v_{ij} \middle| j \in J\right), \left(\min_i v_{ij} \middle| j \in J^-\right), i = 1, 2, \ldots, m \right\}, \\ &= \left\{ v_1^*, v_2^*, \ldots, v_n^* \right\}, \end{aligned} \tag{5.25}$$

$$A^- = \left\{ \left(\min_i v_{ij} \middle| j \in J \right), \left(\max_i v_{ij} \middle| j \in J^- \right), i = 1, 2, \ldots, m \right\}, \\ = \left\{ v_1^-, v_2^-, \ldots, v_n^- \right\}, \tag{5.26}$$

where J is a subset of $\{j = 1, 2, \ldots, n\}$, which presents the benefit criteria (maximum value), whereas J^- is the complement set of J, which refers to the cost criteria (minimum value). The ideal solution (A^*) presents the most preferable alternative. The negative ideal solution (A^-) presents the least preferable alternative.

Step 4: Computing the separation measure

The separation distance between each alternative in V and the ideal and negative ideal solutions are computed using the n-dimension Euclidean distance method. The distance of an alternative from the ideal solution is given by

$$S_i^* = \sqrt{\sum_{j=1}^{n} \left(v_{ij} - v_j^* \right)^2}, \text{ for } i = 1, 2, \ldots, m \tag{5.27}$$

where S_i^* refers to the distance of the ith alternative from the ideal solution. Similarly, the distance of an alternative in V from the negative ideal solution is computed by

$$S_i^- = \sqrt{\sum_{j=1}^{n} \left(v_{ij} - v_j^- \right)^2}, \text{ for } i = 1, 2, \ldots, m \tag{5.28}$$

where, S_i^- refers to the distance of the ith alternative from the negative ideal solution. Each alternative in V has two values, namely S_i^* and S_i^-, at the end of step 4.

Step 5: Computing the relative closeness to the ideal solution.

The relative closeness of the ith alternative (A_i) with respect to the A^* is computed by.

$$C_i^* = \frac{S_i^-}{S_i^- + S_i^*}, \text{ where } i = 1, 2, \ldots, m \text{ and } 0 \leq C_i^* \leq 1 \tag{5.29}$$

It is worth mentioning that $C_i^* = 1$ if and only if $A_i = A^*$, and $C_i^* = 0$ if and only if $A_i = A^-$.

Step 6: Ranking the alternative based on the closeness to the ideal solution.

The set of the alternatives A_i in V are ranked according to the descending order of C_i^*. Therefore, the best (optimal) alternative is the one that has the shortest distance to the A^* and the longest distance from the A^-.

5.4.2 Sizing example

In this example, the system is proposed to supply water for irrigation and drinking for a small village in Kuala Lumbar, Malaysia, with a 30 m^3 daily water demand. The size and performance of the system are tested using an hourly meteorological dataset for 1 year. An iterative numerical program is used to specify the performance of the system for each configuration to constitute the decision matrix (*DM*) used to sort the configurations based on the suggested weights by the experts (developers). The performance of various configurations is represented in *DM* via three criteria: reliability (*LLP*), cost (*LCC*), and excess water (Q_e).

The correlation between the loss of load probability (*LLP*), *LCC*, and the size of the storage tank (C_n) of the system for 20 PV modules with an *LLP* less than or equal to 0.01 is shown in Fig. 5.24. The inverse relationships between LLP and LCC and between LLP and C_n are clear in Fig. 5.24.

Due to the large size of the data, the results of scoring the configuration of the PVPS are presented for the first 11 cases for each developer as tabulated in Tables 5.5–5.7, respectively. The first and second columns of each table represent the number of PV modules and strings connected in series and parallel, respectively. The third column refers to the maximum capacity of the storage tank. The fourth and fifth columns of each developer consider the separation measurements of each alternative (configuration) of weighted normalization *DM* from the ideal (A^+) and negative ideal (A^-) solutions, respectively. The last column of the above tables represents the score of closeness to the ideal solution. According to Tables 5.5–5.7, the PV array with a configuration of five PV modules connected in series and four PV strings connected in parallel is more effective to realize an acceptable balance between the reliability and cost of the system as well as the excess water. The hybrid of the AHP-TOPSIS multicriteria decision-making methods leads to the PVPS configuration $N_s = 5$, $N_p = 4$, and $C_n = 79$ m^3 is the closest configuration to the ideal solution based on the weights of all developers. The closeness scores to the ideal solution of the aforementioned configuration are 0.98539, 0.98419, and 0.98117 for the first, second, and third developers, respectively. It is worth mentioning that the configurations and their closeness to the ideal solution for all developers are consistent. This is due to the weights of criteria of developers being very close to each other, especially for the first and second developers. The weights of *LLP*, *LCC*, and Q_e for the first and second experts are 0.61, 0.09, 0.3, 0.58, 0.11, and 0.31, respectively. The sorting of configurations based on closeness to the ideal solution of the

Table 5.5 Scores based on integrated AHP-TOPSIS for first developer.

N_s	N_p	C_n	S_i^*	S_i^-	C_i^*
5	4	79	0.00172	0.11613	0.98539
5	4	77	0.00286	0.11515	0.97573
5	4	75	0.00434	0.11401	0.96334
5	4	74	0.00564	0.11304	0.95244
5	4	73	0.00696	0.11208	0.94152
5	4	70	0.01095	0.10924	0.90893
5	4	69	0.01228	0.10831	0.89819
5	4	68	0.01361	0.10739	0.88751
5	4	67	0.01495	0.10648	0.87691
5	5	41	0.01591	0.10617	0.86970
5	5	40	0.01597	0.10563	0.86863

Table 5.6 Scores based on integrated AHP-TOPSIS for second developer.

N_s	N_p	C_n	S_i^*	S_i^-	C_i^*
5	4	79	0.00185	0.11500	0.98419
5	4	77	0.00285	0.11409	0.97566
5	4	75	0.00420	0.11305	0.96415
5	4	74	0.00542	0.11217	0.95387
5	4	73	0.00667	0.11129	0.94349
5	4	70	0.01043	0.10871	0.91243
5	4	69	0.01170	0.10786	0.90217
5	4	68	0.01296	0.10703	0.89198
5	4	67	0.01423	0.10620	0.88185
5	4	66	0.01550	0.10538	0.87181
5	5	41	0.01644	0.10418	0.86369

Table 5.7 Scores based on integrated AHP-TOPSIS for third developer.

N_s	N_p	C_n	S_i^*	S_i^-	C_i^*
5	4	79	0.00216	0.11267	0.98117
5	4	77	0.00364	0.11101	0.96826
5	4	75	0.00553	0.10909	0.95172
5	5	41	0.00610	0.11157	0.94817
5	5	40	0.00635	0.11073	0.94574
5	4	74	0.00721	0.10745	0.93715
5	5	37	0.00850	0.10702	0.92644
5	4	73	0.00889	0.10580	0.92246
5	6	68	0.01136	0.11087	0.90703
5	5	35	0.01244	0.10230	0.89156
5	4	70	0.01399	0.10088	0.87821

third expert in Table 5.7 slightly differs from Table 5.5 and 5.6 because in this case, the *LLP* is given the highest weight as compared to other criteria. The average scores of closeness to the ideal solution based on the integrated AHP-TOPSIS for all developers with PVPS configurations are tabulated in Table 5.8. The configuration $N_s=5$, $N_p=4$, and $C_n=79\ m^3$ still has the highest closeness score to the ideal solution, around 0.98358. The above configuration achieves *LLP*, *LCC*, and Q_e of about 0.00041, 10,524.9 USD, and $3422.5 m^3$.

The performance of the PVPS with configuration $N_s=5$, $N_p=4$, and $C_n=79\ m^3$ based on hourly solar irradiance and ambient temperature for 1 year is illustrated in Fig. 5.25. The first and second subplots of Fig. 5.25 illustrate the daily solar irradiance and daily output power of the PV array for a year with average solar irradiance and PV output power over 1 year of around $4349.859\ W/m^2$ and 497.604 W, respectively. The water demand and water flow rate are explained in subplot three of Fig. 5.25, where the red curve refers to the volume of daily water demand. According to subplot three, the system often meets the water demand within most days of the year, except only 2 days, as illustrated in subplots four and six of Fig. 5.25. The last subplots refer to the daily volume of deficit water and the daily loss of load probability. The proposed system offers an *LLP* value during January 1 around 0.1446 with a deficit water volume of $4.338\ m^3$. In addition, the PVPS offers *LLP* on December 25 around 0.0042 with $0.1254\ m^3$ as the deficit water volume. The proposed system offers high reliability with the *LLP* and deficit water being zero during the year, except for January 1 and

Table 5.8 Average scores based on integrated AHP-TOPSIS for all developers.

N	N_s	N_p	C_n	*LLP*	*LCC*	Q_e	Average C_i^*
20	5	4	79	0.00041	10,524.9	3422.5	0.98358
20	5	4	77	0.00059	10,479.4	3426.5	0.97322
20	5	4	75	0.00080	10,433.9	3430.9	0.95974
20	5	4	74	0.00099	10,411.2	3433.9	0.94782
20	5	4	73	0.00117	10,388.4	3436.9	0.93582
20	5	4	70	0.00172	10,320.1	3445.9	0.89986
25	5	5	41	0.00038	10,342.9	7141.5	0.89385
25	5	5	40	0.00047	10,320.1	7143.5	0.89235
20	5	4	69	0.00190	10,297.4	3448.8	0.88794
25	5	5	37	0.00088	10,251.9	7151.0	0.88135
20	5	4	68	0.00208	10,274.6	3451.9	0.87606

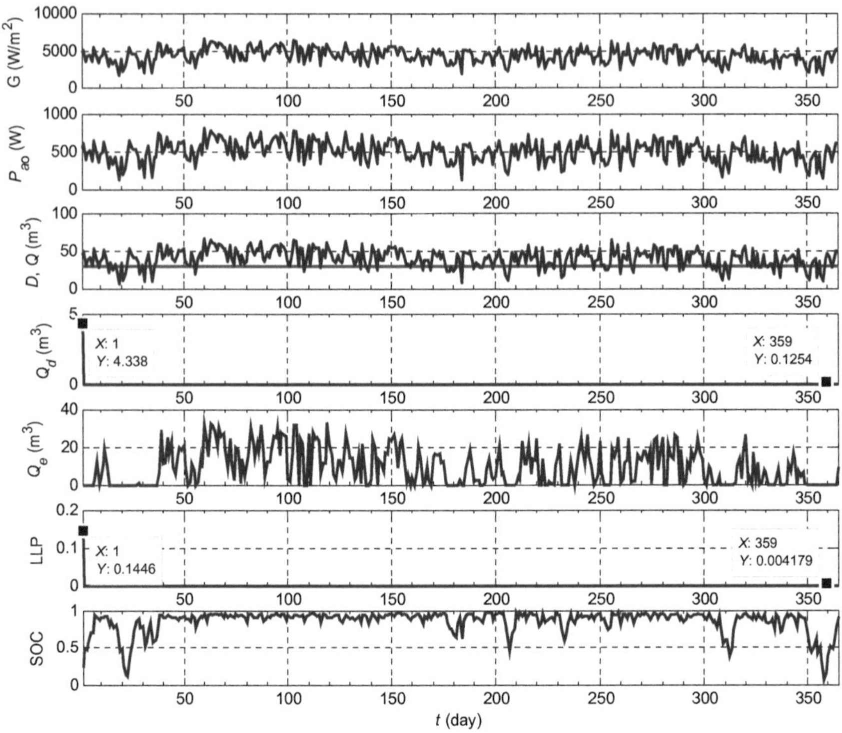

Fig. 5.25 Daily performance of system with configuration is $N_S=5$, $N_P=4$, $C_n=79\ m^3$.

December 25, as shown in the fourth and sixth subplots of Fig. 5.25. The hourly excess water volume of the proposed PVPS is accumulated daily over a year and is illustrated in the fifth subplot of Fig. 5.25. The last subplot of Fig. 5.25 shows the daily state of charge of the water storage tank over a year. The trend of the daily state of charge of the storage tank is often higher than 50% during most days of the year, which is due to the high productivity of the proposed PVPS.

The average daily solar irradiance, water flow rate, volume of water deficit, *LLP*, *SOC*, and volume of resident water of the storage tank for shortage days during the year are tabulated in Table 5.9. The minimum shortage occurred on December 25 when the average daily solar irradiance, Q, Q_d, *LLP*, *SOC*, and C_{res} were $4174.8\ W/m^2$, $38.3579\ m^3/day$, $0.1254 m^3$, 0.0042, 0.1679, and $13.2615\ m^3$, respectively. On the other hand, the maximum shortage occurred on January 1 when the average daily

Table 5.9 Performance of system during shortage days.

Date	G (W/m^2)	Q (m^3/day)	Q_d (m^3)	LLP	SOC	C_{res} (m^3)
1-Jan	4663.8	44.3541	4.3376	0.1446	0.2366	18.6916
25-Dec	4174.8	38.3579	0.1254	0.0042	0.1679	13.2615

solar irradiance, Q, Q_d, LLP, SOC, and C_{res} were 4663.8 W/m^2, 44.3541 m^3/day, 4.3376m^3, 0.1446, 0.2366, and 18.6916 m^3, respectively. However, the system LLP and water deficit volume over a year are 0.0004 and 4.4629 m^3, respectively. Although the daily water flow rate on January 1 and December 25 exceeded the daily water demand, the LLP of the aforementioned days was not zero. Actually, the shortage particularly occurred in the early hours of the morning because the low solar irradiance as well as the resident water in the storage tank were very little before starting these days.

The monthly performance of the proposed PVPS is tabulated in Table 5.10. The hourly meteorological data for 1 year were collected at the Subung meteorological station in Subung Jaya Klang Valley, Malaysia. The daily average monthly solar irradiance and ambient temperature are depicted in the second and third columns, where the first column refers to the month sequence. The maximum and minimum ambient temperatures are registered in March (30.077°C) and November (27.666°C), respectively. The highest productivity of the system occurred in March when the water flow rate and daily average monthly solar irradiance were 1517.553m^3 and 5131.552 W/m^2, respectively. On the other hand, the lowest productivity occurred in November with a water flow rate around 1016.310m^3 at a daily average monthly solar irradiance of 3869.177 W/m^2. However, the maximum shortage over the year occurred in January when the monthly Q_d and LLP values were 4.338 m^3 and 0.00466, respectively. By contrast, the minimum shortage over the year occurred in December with monthly Q_d and LLP values of 0.125 m^3 and 0.00013, respectively. Finally, the monthly state of charge of the storage tank was always more than 50% over the year, as illustrated in the ninth column of Table 5.10.

The following MATLAB code illustrates the executed example for the multicriteria decision-making method. The sizing code of the system itself is the code for the numerical sizing method provided in Chapter 4.

Table 5.10 Monthly performance of PVPS with configuration is $N_s = 5$, $N_p = 4$, and $C_n = 79\ m^3$.

Month	G (W/m^2)	T (°C)	Q (m^3)	D (m^3)	Q_e (m^3)	Q_d (m^3)	LLP	SOC	C_{res} (m^3)
1	3817.458	28.963	1025.078	930	59.050	4.338	0.00466	0.511	40.365
2	4434.214	28.934	1139.535	840	268.397	0.000	0.00000	0.905	71.503
3	5131.552	30.077	1517.553	930	586.052	0.000	0.00000	0.924	73.004
4	5054.807	29.233	1445.611	900	546.662	0.000	0.00000	0.911	71.953
5	4573.365	29.919	1311.601	930	379.917	0.000	0.00000	0.932	73.637
6	4207.677	29.527	1121.575	900	245.212	0.000	0.00000	0.633	50.000
7	4064.742	28.917	1098.877	930	142.186	0.000	0.00000	0.971	76.691
8	4308.677	28.872	1215.402	930	288.838	0.000	0.00000	0.927	73.255
9	4414.053	28.131	1213.969	900	313.147	0.000	0.00000	0.938	74.077
10	4542.719	28.323	1303.804	930	381.316	0.000	0.00000	0.843	66.564
11	3869.177	27.666	1016.310	900	125.140	0.000	0.00000	0.731	57.735
12	3779.865	27.777	1030.521	930	86.615	0.125	0.00013	0.908	71.767

```
%%%%%%%MCDM for PVPS based on LLP, LCC and Qe criterias
%%%%%%%TOPSIS Script code for 3-Critera
%%%%%%%Created on 10/12/2019
clc;clear all;close all;
t1=cputime;
N=xlsread('Result_Sizing_Numerical_Iterative_Method.xlsx',2,'B4:
   B38685');
Ns=xlsread('Result_Sizing_Numerical_Iterative_Method.xlsx',2,'C4:
   C38685');
Np=xlsread('Result_Sizing_Numerical_Iterative_Method.xlsx',2,'D4:
   D38685');
Cn=xlsread('Result_Sizing_Numerical_Iterative_Method.xlsx',2,'E4:
   E38685');
LLP=xlsread('Result_Sizing_Numerical_Iterative_Method.xlsx',2,'F4:
   F38685');
LCC=xlsread('Result_Sizing_Numerical_Iterative_Method.xlsx',2,'G4:
   G38685');
Qe=xlsread('Result_Sizing_Numerical_Iterative_Method.xlsx',2,'I4:
   I38685');
Cost_Year=xlsread('Result_Sizing_Numerical_Iterative_Method.
   xlsx',2,'H4:H38685');
Qd=xlsread('Result_Sizing_Numerical_Iterative_Method.xlsx',2,'J4:
   J38685');
COU=xlsread('Result_Sizing_Numerical_Iterative_Method.xlsx',2,'K4:
   K38685');
m=length(LLP)
%%%%%%%%%Normalization%%%%%%%%%%%%
LLPND=sqrt(sum(LLP.^2));
LCCND=sqrt(sum(LCC.^2));
QeND=sqrt(sum(Qe.^2));
LLPN=LLP./LLPND;
LCCN=LCC./LCCND;
QeN=Qe./QeND;
R=[LLPN,LCCN,QeN];
R1=LLPN;
R2=LCCN;
R3=QeN;
%%%%%%%Declaration of weightes%%%%%%
w1=0.6;
w2=0.1;
w3=0.3;
%%%%%%Construction the weighted normalized decision matrix%%%%%%
```

Continued

—cont'd

```
V1=w1.*R(:,1);
V2=w2.*R(:,2);
V3=w3.*R(:,3);
V=[V1,V2,V3];
%%%%%%%Computing the positive and negative ideal solutions%%%%%%
V1OP1=min(V1);
V2OP2=min(V2);
V3OP3=min(V3);
V1ON1=max(V1);
V2ON2=max(V2);
V3ON3=max(V3);
VOP=[V1OP1,V2OP2,V3OP3];
VON=[V1ON1,V2ON2,V3ON3];
%%%%%%%%%%%%%%%%%%%%%%%%%%%%%%
V1OP=zeros(m,1);
V2OP=zeros(m,1);
V3OP=zeros(m,1);
V1ON=zeros(m,1);
V2ON=zeros(m,1);
V3ON=zeros(m,1);
%%%%%%%%%%%%%%%%%%%%%%%%%%%%%%%
V1OP(:,1)=V1OP1;
V2OP(:,1)=V2OP2;
V3OP(:,1)=V3OP3;
V1ON(:,1)=V1ON1;
V2ON(:,1)=V2ON2;
V3ON(:,1)=V3ON3;
%%%%%%%%%Separation measure calculation%%%%%%%%%%%%%%%%%%%%%%%%%%
%%%%%%%%%Euclidean distance from positive ideal solution%%%%%%%%%%
C1P=(V1-V1OP).^2;
C2P=(V2-V2OP).^2;
C3P=(V3-V3OP).^2;
S1P=C1P+C2P+C3P;
%%%%%%%%%Euclidean distance from negative ideal solution%%%%%%%%%%
C1N=(V1-V1ON).^2;
C2N=(V2-V2ON).^2;
C3N=(V3-V3ON).^2;
S1N=C1N+C2N+C3N;
SN=(S1N).^(0.5);
C=SN./(SP+SN);
%%%%%%%%%%%%%%%%%%%%%%%%%%%%%Ascending sorting%%%%%%%%%%%%%%%%%%%%
[CI, ID]=sort(C,'ascend');
```

—cont'd

```
LLP=LLP(ID);
LCC=LCC(ID);
Qe=Qe(ID);
Cost_Year=Cost_Year(ID);
Qd=Qd(ID);
COU=COU(ID);
N=N(ID);
Ns=Ns(ID);
Np=Np(ID);
Cn=Cn(ID);
V1=V1(ID);
V2=V2(ID);
V3=V3(ID);
V1OP=V1OP(ID);
V2OP=V2OP(ID);
V3OP=V3OP(ID);
V1ON=V1ON(ID);
V2ON=V2ON(ID);
V3ON=V3ON(ID);
SP=SP(ID);
SN=SN(ID);
R1=R1(ID);
R2=R2(ID);
R3=R3(ID);
t=cputime t1
%%%%%%%%%%%%%Writing the results in Excel sheet%%%%%%%%%%%%%%%%%%%%
xlswrite('Result_Sizing_Numerical_Iterative_Method.xlsx',N,2,'N4');
xlswrite('Result_Sizing_Numerical_Iterative_Method.xlsx',Ns,2,'O4');
xlswrite('Result_Sizing_Numerical_Iterative_Method.xlsx',Np,2,'P4');
xlswrite('Result_Sizing_Numerical_Iterative_Method.xlsx',Cn,2,'Q4');
xlswrite('Result_Sizing_Numerical_Iterative_Method.xlsx',LLP,2,'R4');
xlswrite('Result_Sizing_Numerical_Iterative_Method.xlsx',LCC,2,'S4');
xlswrite('Result_Sizing_Numerical_Iterative_Method.xlsx',
   Cost_Year,2,'T4');
xlswrite('Result_Sizing_Numerical_Iterative_Method.xlsx',Qe,2,'U4');
xlswrite('Result_Sizing_Numerical_Iterative_Method.xlsx',Qd,2,'V4');
xlswrite('Result_Sizing_Numerical_Iterative_Method.xlsx',COU,2,'W4');
xlswrite('Result_Sizing_Numerical_Iterative_Method.xlsx',R1,2,'X4');
xlswrite('Result_Sizing_Numerical_Iterative_Method.xlsx',R2,2,'Y4');
xlswrite('Result_Sizing_Numerical_Iterative_Method.xlsx',R3,2,'Z4');
xlswrite('Result_Sizing_Numerical_Iterative_Method.xlsx',V1,2,'AA4');
```

Continued

—cont'd

```
xlswrite('Result_Sizing_Numerical_Iterative_Method.xlsx',V2,2,'AB4');
xlswrite('Result_Sizing_Numerical_Iterative_Method.xlsx',V3,2,'AC4');
xlswrite('Result_Sizing_Numerical_Iterative_Method.xlsx',
   V10P,2,'AD4');
xlswrite('Result_Sizing_Numerical_Iterative_Method.xlsx',
   V20P,2,'AE4');
xlswrite('Result_Sizing_Numerical_Iterative_Method.xlsx',
   V30P,2,'AF4');
xlswrite('Result_Sizing_Numerical_Iterative_Method.xlsx',
   V10N,2,'AG4');
xlswrite('Result_Sizing_Numerical_Iterative_Method.xlsx',
   V20N,2,'AH4');
xlswrite('Result_Sizing_Numerical_Iterative_Method.xlsx',
   V30N,2,'AI4');
xlswrite('Result_Sizing_Numerical_Iterative_Method.xlsx',SP,2,'AJ4');
xlswrite('Result_Sizing_Numerical_Iterative_Method.xlsx',SN,2,'AK4');
xlswrite('Result_Sizing_Numerical_Iterative_Method.xlsx',CI,2,'AL4');
xlswrite('Result_Sizing_Numerical_Iterative_Method.xlsx',t,2,'AM4');
```

CHAPTER 6

A feasibility and load sensitivity analysis of a photovoltaic water pumping system with a battery and a diesel generator

Contents

6.1 Introduction

In general, most of designing methods of photovoltaic water pumping systems (PVPS) consider storage tanks as a storage unit. However, some research works are devoted to combining a battery with PVPS. On the other hand, other research focuses on the feasibility of using a PVPS versus pumping system that is based on diesel generator (DG).

As a fact, PVPS still suffers from the most important challenge, which is the mismatch between the PV-generated energy and the load demand. This is because of the intermittent nature of solar energy. Therefore, the reliability of such a system is critical and backup energy sources, such as diesel generator (DG) or storage element such as a battery, are needed.

In this chapter, a feasibility study and a load sensitivity analysis of PVPS and PV/DG/battery hybrid pumping systems are presented. The analysis is based on technoeconomic aspects to select a pumping system design from

Photovoltaic Water Pumping Systems
https://doi.org/10.1016/B978-0-12-821231-8.00006-8

various proposed systems in order to minimize the life cycle cost of system with 100% reliability ($LLP = 0$). An energy balance computation is done for each hour of a year to obtain a perfect design for a pumping system for a certain load at a specific site.

The PVPSs in this chapter are designed using methods discussed in Chapter 5 and using system models discussed in Chapter 3.

6.2 PVPS with a storage battery

One of the possible solutions to improve the operation stability of PVPS is to integrate another storage element such as battery with the system. In general, the storage tank is usually a better option compared to battery in PVPS due to its simplicity, lower cost, less need for maintenance, and longer life. However, battery can be used for strict reliable requirements. This is because the lead-acid batteries have acceptable life cycle costs and technical performance characteristics, they are currently being used as energy storage elements for PV system applications. The capacity of battery mainly depends on the hourly water demand and the time period that is required to cover the water demand from the battery bank when the solar irradiance is absent, or when the PV-generated power is smaller than the demand. The maximum capacity of the battery bank can be calculated by

$$C_B = \frac{E_{mmin} A_H}{\zeta_B DOD} \tag{6.1}$$

where E_{mmin} is the hourly motor input power required to pump the water demand (W), A_H is the autonomy hours in the maximum shortage day (h), ζ_B is the efficiency of the battery, and DOD is the depth of discharge of the battery bank. It is worth mentioning that the lifespan of the battery bank mainly depends on the DOD and the number of charge/discharge cycles. The DOD can be written as

$$DOD = 1 - SOC_{Bmin} \tag{6.2}$$

The autonomy hours represents the number of hours for a day that has the maximum deficit within a year, and can be written as

$$A_H = \frac{\text{Max} \sum_{t=1}^{12} \text{hourly deficit water} \left(Q_d(t) \right)}{\text{Hourly water demand} \left(2.5\,\text{m}^3/\text{h} \right)} \tag{6.3}$$

According to the example in Section 5.3, the maximum deficit occurred on December 24 with a deficit water volume of 21.269 m^3. Thus, the total number of deficit hours (A_H) is 8.508 h. According to the performance of the PVPS, the value of E_{mmin} is found to be 324 W (V_m = 60 V and I_m = 5.4 A), which enables the system to pump water around 2.5 m^3/h. Assume the $\zeta_B$and the DOD are chosen to be 85% and 80%, respectively. Then, the maximum capacity of the battery bank is assumed to be 4.5 kWh.

6.2.1 Operation scenario of a PVPS with a battery

In this chapter, three operation modes are considered when combining a battery with a PVPS: normal, charging, and discharging. The switching between these modes is based on the output power of the PV array, considering ($E(t)$), $SOC_T(t)$, and $SOC_B(t)$. The $E(t)$ is compared with the minimum input power of the motor (E_{mmin}) to produce the desired water demand. When $E(t)$ is greater than E_{mmin} and SOC_T $(t-1)$ and SOC_B $(t-1)$ is 100%, the excess energy is utilized to charge the battery bank (charging mode). Otherwise, the excess energy is converted to excess water (normal mode). In addition, the system is operated in normal mode when the $E(t)$ is less than E_{mmin}, and $SOC_T(t-1)$ is greater than or equal to SOC_{Tmin}(SOC_{Tmin}, equals 2.5 m^3/C_n). Finally, the system is operated in a discharging mode when the $E(t)$ is less than E_{mmin} and the $SOC_T(t-1)$ is less than SOC_{Tmin}. In discharging mode, the battery is discharged by an energy amount equal to the difference between $E(t)$ and E_{mmin} to enable the system to pump the volume of the demanded water. Fig. 6.1 shows the operation scenario of the proposed PVPS with a battery storage element. The main sources for fulfilling the water demand are the PV array and the storage tank. The battery bank is represented as a backup source for energy when the primary sources fail to fulfill the water demand. The proposed strategy of operation modes assists in prolonging the battery life, and consequently reducing the life cycle cost and the cost of the unit of water. Furthermore, charging the storage tank by water has a higher priority than charging the battery, which is to reduce the number of charge/discharge cycles and consequently increase the lifespan of the battery bank.

6.3 PVPS with diesel generator

A diesel generator in a hybrid system plays a role as a backup energy source when the PV array is unable to meet the load demand. The combination of a

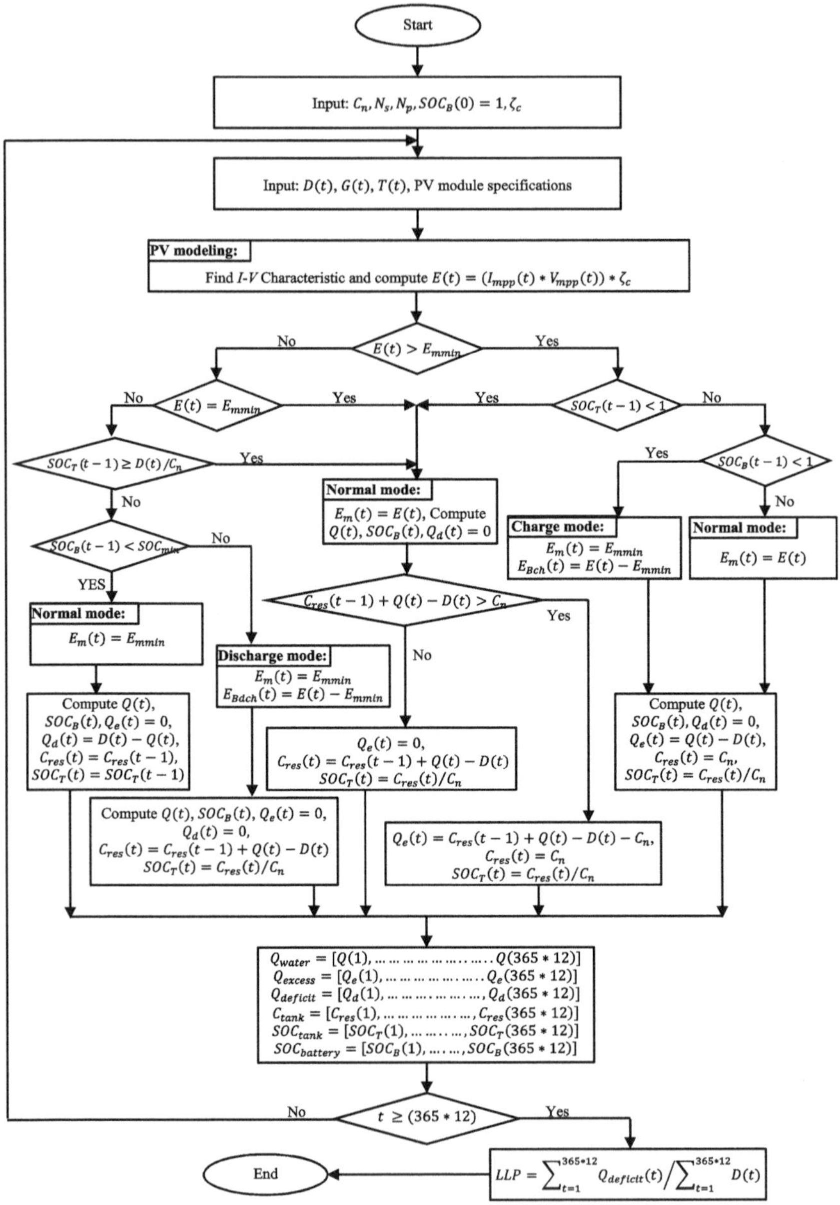

Fig. 6.1 Operation modes of PVPS with battery.

PV system with a DG to power the pumping system helps to save fuel as well as makes the operation of the system more stable.

The amount of fuel consumption of the DG mainly depends on the rated power and the output power of the DG. The best way to compute the fuel consumption of the DG at any load is by using the manufacturer fuel consumption curve of the DG. In general, the manufacturers usually provide only the value of fuel consumption at the rated power. The methodology that is used to compute the fuel consumption is described below

$$F_{DGC} = A_{DG} P_{DG} + B_{DG} P_{DGr} \tag{6.4}$$

where F_{DGC} is the fuel consumption by DG at a given time (L/h), P_{DGr} and P_{DG} are the rated and instantaneous output power of DG (kW), respectively, and A_{DG} and B_{DG} are the coefficients of the DG fuel consumption curve (L/kWh). Typically, the values of A_{DG} and B_{DG} are close to 0.246 L/kWh and 0.08145 L/kWh, respectively. The daily fuel consumption (FC) can be computed by

$$FC = \sum_{t=1}^{12} F_{DGC} \tag{6.5}$$

The DG emissions include CO_2, CO, SO_2, and NO_x gases based on the energy generated. CO_2 represents the largest amount of gas emitted from the DG when the fuel is burnt. Therefore, the measure of pollutants released from the DG is represented by the kg of emitted CO_2. The amount of emitted CO_2 depends on the DG fuel characteristics, and it belongs to the range of [2.4, 2.8] kg/L. The total CO_2 that is emitted during an hour, CO_{2_hour}, can be computed by

$$CO_{2\_hour} = F_{DGC} * CO_{2\_litre} \tag{6.6}$$

where CO_{2_litre} is the amount of kg of CO_2 emitted per liter of fuel, and CO_{2_hour} is the amount of kg of CO_2 emitted in an hour. The mass of CO_2 emitted per day (CO_2) is calculated by

$$CO_2 = \sum_{t=1}^{12} CO_{2\_hour} \tag{6.7}$$

6.3.1 Operation scenario of PVPS with DG

There are two operation modes of a system powered by a PV array and a DG: the PV standalone operation mode and the DG operation mode. The switching between them depends on the output power of the PV array, considering ($E(t)$) and $SOC_T(t)$ at a given time. When the $SOC_T(t-1)$ is greater than or equal to SOC_{Tmin}, (regardless of the value of $E(t)$), the system will be operated in PV standalone mode. Moreover, the system is operated in PV standalone mode when $E(t)$ is greater than or equal *to* E_{mmin}, (regardless of the status of $SOC_T(t-1)$). The system is operated in DG mode when

$E(t)$ and $SOC_T(t-1)$ are less than E_{mmin} and SOC_{Tmin}, respectively. In DG mode, the DG is worked and the PV array is disconnected from system. The PV standalone and DG operation modes of the proposed system are shown in Fig. 6.2.

6.4 Sensitivity analysis of PVPS performance with different back sources

The results of a technoeconomic feasibility study and a load sensitivity analysis of a standalone DG-battery hybrid pumping system are presented in this section. Table 6.1 shows the various proposed scenarios and their system configurations. Scenario #1 refers to the configuration obtained from the example in Section 5.3 (PV system with a storage tank). Scenario #2 is similar to Scenario #1, but the initial status of the storage tank is assumed to be full with water. In Scenario #3, a lead acid battery bank with a capacity (C_B) of 4.5 kWh is used to substitute the deficit energy of the PV array during shortage days. In Scenario #4, a diesel generator with a capacity (C_{DG}) of 1.5 kVA is merged with a PVPS to fulfill the load demand when there is insufficient energy from the PV array. The size of the storage tank is assumed to be 80 m^3 in Scenario #5 (initial empty tank) while an initial full tank is assumed for the sixth scenario (Scenario #6). The last scenario (Scenario #7) is a pumping system powered by a DG only.

6.4.1 Technical analysis

In this section a technical analysis is conducted for various scenarios. The analysis is conducted based on the annual hourly meteorological data used in Section 5.3. The average hourly pumping water volume by PVPS is 3.297 m^3 with the *LLP* about 0.005024 for a year's time. The performance of Scenario #1 for specific (shortage) days is briefly tabulated in Table 6.2 for comparison purposes. The maximum deficit occurs on December 24 with a daily *LLP* and Q_d of 0.709 and 21.269 m^3, respectively. On the other hand, the minimum deficit occurs on Nov. 9, where the daily *LLP* and Q_d are 0.007 and 0.223 m^3, respectively.

The proposed Scenario #2 of PVPS is similar to scenario #1, except the storage tank is initially full with water to overcome the shortage in the first operational hours of PVPS. Therefore, the shortage days in this scenario are only eight days. The daily performance of PVPS (Scenario #2) during shortage days is illustrated in Table 6.2. The system pumps a 3.297 m^3 as average daily water volume with *LLP* is around 0.004628 for a year time. *LLP* and

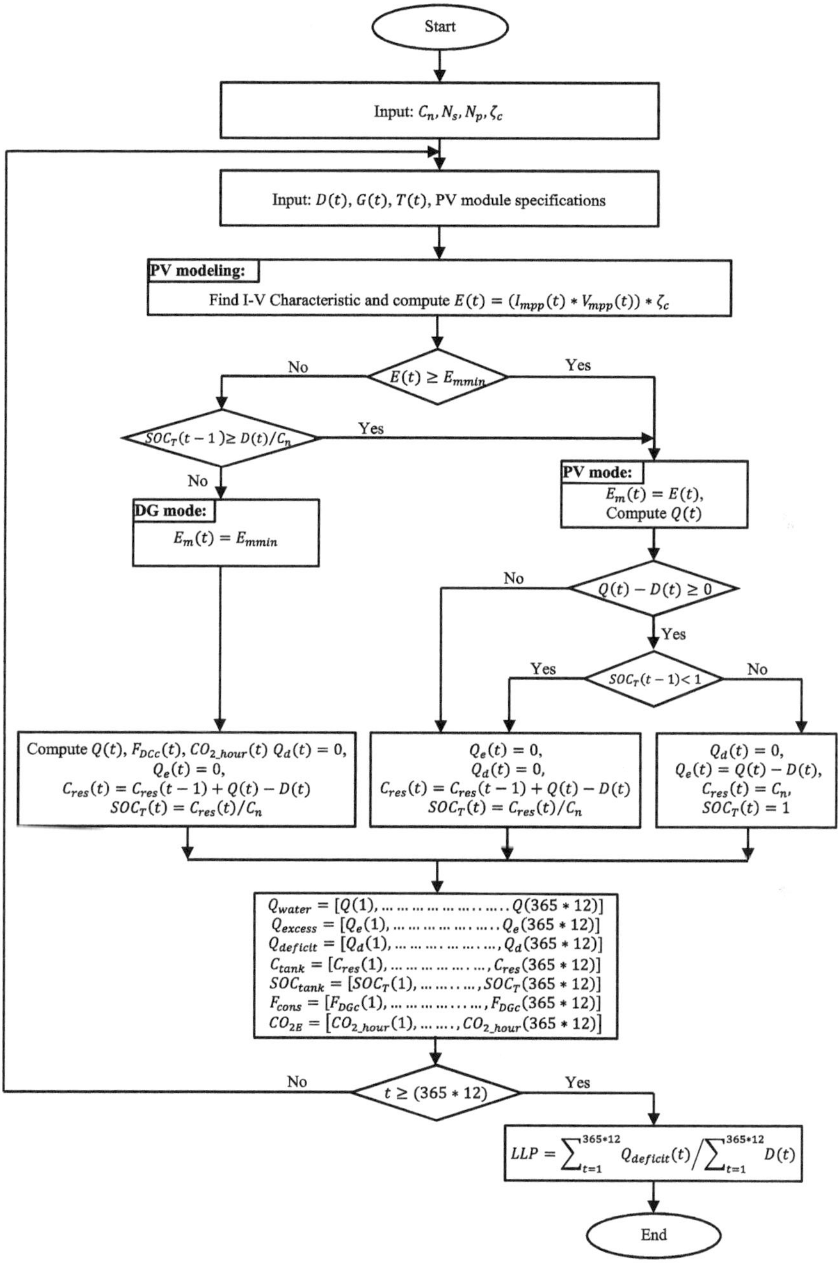

Fig. 6.2 PV standalone and DG operation modes of PVPS with DG.

Table 6.1 Various scenarios and their system configurations.

Scenario	N_s	N_p	C_n (m^3)	Initial tank status	C_B (kWh)	C_{DG} (kVA)
1	5	4	52	Empty	0	0
2	5	4	52	Full	0	0
3	5	4	52	Empty	4.5	0
4	5	4	52	Empty	0	1.5
5	5	4	80	Empty	0	0
6	5	4	80	Full	0	0
7	0	0	30	Empty	0	1.5

Q_d are zero in the first day of January Due to the status of storage tank (initially full). Maximum deficit still occurs on December 24, due to the water level in the storage tank for previous day (December 23) was zero and the daily pumped water volume is 8.732 m^3. The maximum daily *LLP* and Q_d are 0.709 and 21.269 m^3, respectively. On the other hand, the minimum deficit occurs on November 9, where the daily *LLP* and Q_d are 0.007 and 0.223 m^3, respectively.

In Scenario #3, two storage elements are used. The first and main one is a 52 m^3 empty storage tank while the second one is a 4.5 kWh battery bank. As stated in Section 3.3, the capacity of the battery bank is chosen based on the autonomy (deficit) hours in the maximum shortage day, where the battery is discharged in deficit hours only. The technical daily performance of Scenario #3 of a PVPS for 1 year is shown in Fig. 6.3. The daily solar irradiance and PV array power production are illustrated in the first and second subplots of Fig. 6.3. The collected daily solar irradiance and the average daily output power of the PV array are 4349.859 W/m^2 and 497.604 W, respectively. The third subplot of Fig. 6.3 indicates the daily water flow rate of the proposed system and the water demand (*D*). The system pumps an average hourly water volume around 3.293 m^3, and the system always meet the water demand over the year with daily Q_d and *LLP* values equal to zero, as indicated in the fourth and sixth subplots of Fig. 6.3, respectively. The volume of excess water (Q_e) is illustrated in the fifth subplot of Fig. 6.3 while the daily state of charge of both the tank (SOC_T) and the battery (SOC_B) are shown in the seventh and eighth subplots of Fig. 6.3, respectively. Table 6.3 illustrates the hourly SOC_T and SOC_B of Scenario #3 of the PVPS for specific days that have water shortages with Scenario #1. For the first two hours of January 1, the solar irradiance is very weak and the PV array is unable to produce adequate energy to pump 2.5 m^3 while the storage tank is initially empty. Therefore, the battery will be in discharge mode, where the output

Table 6.2 Performance of various pumping system scenarios within specific (shortage) days.

Date		1/ January	21/ January	22/ January	23/ January	9/ November	22/ December	23/ December	24/ December	25/ December
Daily G_T (Wh/m^2)		4663.8	2163.9	2783.1	3602.9	4197.2	3966.5	2638.9	1591.7	4174.8
Scenario #1	Q (m^3/day)	44.354	11.529	19.792	29.117	37.524	35.555	17.083	8.732	38.358
	Q_d (m^3/day)	4.338	6.656	10.208	6.466	0.223	0.510	0.443	21.269	4.904
	C_{res} (m^3)	18.692	0	0	5.582	11.436	12.474	0	0	13.262
	LLP	0.145	0.222	0.340	0.216	0.007	0.017	0.015	0.709	0.164
Scenario #2	Q (m^3/day)	44.354	11.529	19.792	29.117	37.524	35.555	17.083	8.732	38.358
	Q_d (m^3/day)	0	6.656	10.208	6.466	0.223	0.510	0.443	21.269	4.904
	C_{res} (m^3)	45.029	0	0	5.582	11.436	12.474	0	0	13.262
	LLP	0	0.222	0.340	0.216	0.007	0.017	0.015	0.709	0.164
Scenario #3	Q (m^3/day)	48.702	19.005	31.419	35.597	38.941	37.479	17.083	30.677	43.277
	Q_d (m^3/day)	0	0	0	0	0	0	0	0	0
	C_{res} (m^3)	18.702	0.820	2.239	7.836	12.630	13.889	0.971	1.648	14.925
	LLP	0	0	0	0	0	0	0	0	0
Scenario #4	Q (m^3/day)	48.702	19.005	31.419	35.597	38.941	37.479	17.083	30.677	43.277
	Q_d (m^3/day)	0	0	0	0	0	0	0	0	0
	C_{res} (m^3)	18.702	0.820	2.239	7.836	12.630	13.889	0.971	1.648	14.925
	LLP	0	0	0	0	0	0	0	0	0
	FC (L/day)	0.865	1.297	2.162	1.297	0.432	0.432	0.000	4.756	1.297
	CO_2 (kg)	2.162	3.243	5.405	3.243	1.081	1.081	0	11.890	3.243

Continued

Table 6.2 Performance of various pumping system scenarios within specific (shortage) days—cont'd

Date		1/ January	21/ January	22/ January	23/ January	9/ November	22/ December	23/ December	24/ December	25/ December
Scenario #5	Q (m^3/day)	44.3541	11.529	19.792	29.117	37.524	35.555	17.083	8.732	38.358
	Q_d (m^3/day)	4.338	0	0	0	0	0	0	0	0
	C_{res} (m^3)	18.692	21.344	11.137	10.253	39.213	39.964	27.047	5.778	14.136
	LLP	0.1446	0	0	0	0	0	0	0	0
Scenario #6	Q (m^3/day)	44.354	11.529	19.792	29.117	37.524	35.555	17.083	8.732	38.358
	Q_d (m^3/day)	0	0	0	0	0	0	0	0	0
	C_{res} (m^3)	73.029	21.344	11.137	10.253	39.213	39.964	27.047	5.778	14.136
	LLP	0	0	0	0	0	0	0	0	0
Scenario #7	Q (m^3/day)	30.061	30.061	30.061	30.061	30.061	30.061	30.061	30.061	30.061
	Q_d (m^3/day)	0	0	0	0	0	0	0	0	0
	C_{res} (m^3)	0.061	1.270	1.331	1.391	18.932	21.533	21.593	21.654	21.714
	LLP	0	0	0	0	0	0	0	0	0
	FC (L/day)	5.188	5.188	5.188	5.188	5.188	5.188	5.188	5.188	5.188
	CO_2 (kg)	12.971	12.971	12.971	12.971	12.971	12.971	12.971	12.971	12.971

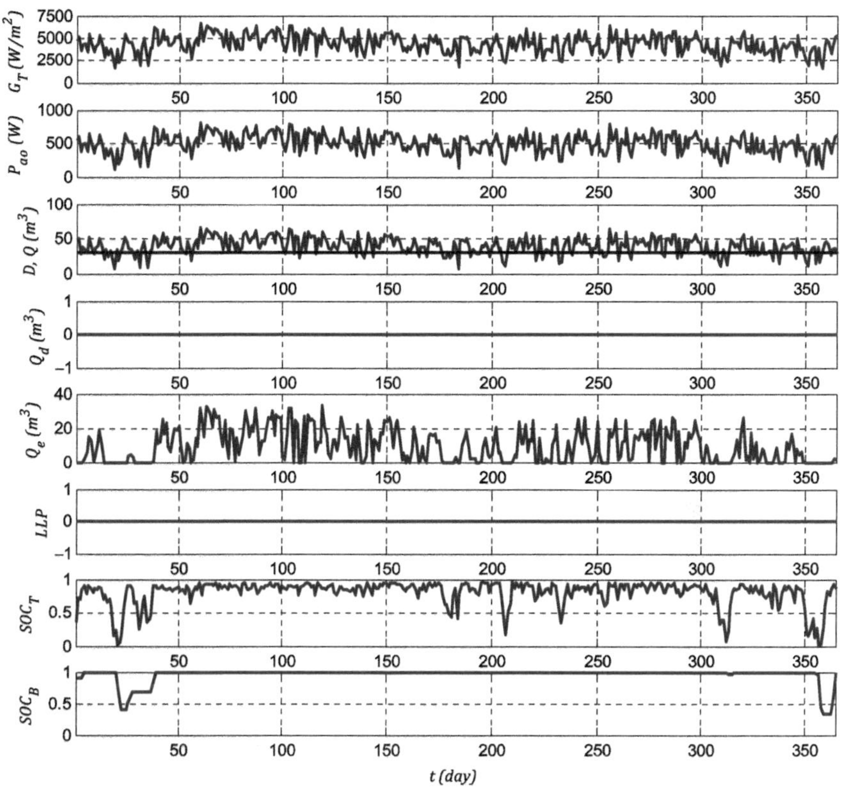

Fig. 6.3 Daily performance of PVPS of Scenario #3 for a year.

energy of the PV array is aggregated with the discharge energy from the battery to pump the water demand. It is worth stating that the battery works in discharge mode for only 8 days during the year.

Table 6.2 shows the daily solar irradiance, water flow rate, deficit water, resident water in the storage tank, and the daily loss of load probability of the PVPS during specific days. The lowest pumping water volume occurred on Jan. 21 and Dec. 23 (less than 20 m^3). It is worth mentioning that the water flow rate is directly related to solar irradiance, but on December 24, a conflict occurred in this relation. This is due to the fact that on December 24, the system solely depends on the PV array, where the battery bank is used as the main source to pump the water demand. According to Table 6.2, the *LLP* and Q_d are zero during shortage days that occurred with Scenario #1 and the level of water is changed within a range of [0.82, 18.702] m^3.

Table 6.3 Hourly SOC_T and SOC_B of PVPS of Scenario #3 for specific days.

	1/January		21/January		22/January		23/January		9/November		22/December		23/December		24/December		25/December	
Hour	SOC_T	SOC_B	SOC_T	SOC_B	SOC_T	SOC_B	SOC_T	SOC_B	SOC_T	SOC_B	SOC_T	SOC_B	SOC_T	SOC_B	SOC_T	SOC_B	SOC_T	SOC_B
7	0.000	0.937	0.179	1.000	0.016	0.753	0.043	0.478	0.023	1.000	0.075	1.000	0.219	0.960	0.019	0.898	0.032	0.379
8	0.000	0.900	0.132	1.000	0.016	0.694	0.043	0.422	0.023	0.972	0.027	1.000	0.174	0.960	0.019	0.835	0.032	0.340
9	0.021	0.900	0.113	1.000	0.016	0.660	0.043	0.391	0.043	0.972	0.027	0.960	0.173	0.960	0.019	0.773	0.032	0.331
10	0.096	0.900	0.112	1.000	0.040	0.660	0.078	0.391	0.104	0.972	0.059	0.960	0.230	0.960	0.019	0.710	0.095	0.331
11	0.201	0.900	0.119	1.000	0.100	0.660	0.100	0.391	0.166	0.972	0.165	0.960	0.193	0.960	0.019	0.647	0.119	0.331
12	0.274	0.900	0.089	1.000	0.108	0.660	0.087	0.391	0.253	0.972	0.262	0.960	0.176	0.960	0.019	0.604	0.196	0.331
13	0.349	0.900	0.076	1.000	0.133	0.660	0.122	0.391	0.295	0.972	0.312	0.960	0.212	0.960	0.019	0.590	0.282	0.331
14	0.430	0.900	0.052	1.000	0.123	0.660	0.177	0.391	0.384	0.972	0.358	0.960	0.167	0.960	0.019	0.580	0.366	0.331
15	0.494	0.900	0.015	1.000	0.089	0.660	0.215	0.391	0.373	0.972	0.362	0.960	0.140	0.960	0.031	0.580	0.412	0.331
16	0.456	0.900	0.016	0.941	0.043	0.660	0.225	0.391	0.339	0.972	0.361	0.960	0.112	0.960	0.031	0.557	0.383	0.331
17	0.408	0.900	0.016	0.878	0.043	0.603	0.199	0.391	0.291	0.972	0.315	0.960	0.067	0.960	0.032	0.504	0.335	0.331
18	0.360	0.900	0.016	0.816	0.043	0.540	0.151	0.391	0.243	0.972	0.267	0.960	0.019	0.960	0.032	0.442	0.287	0.331

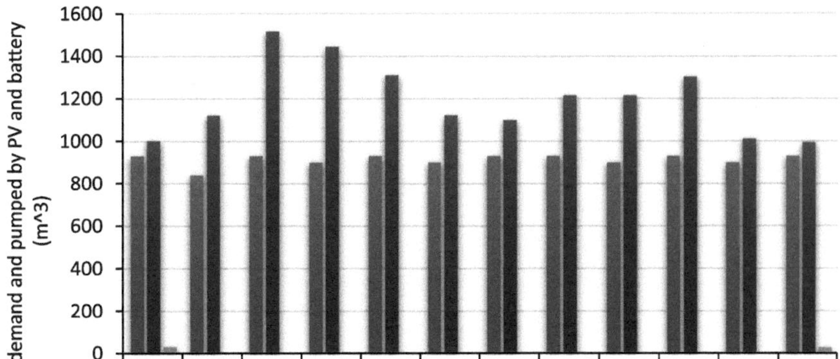

Month	1	2	3	4	5	6	7	8	9	10	11	12
D	930	840	930	900	930	900	930	930	900	930	900	930
PV	1003.4	1122.3	1517.6	1445.6	1311.6	1121.6	1098.9	1215.4	1214.0	1303.8	1012.4	995.8
Battery	29.9	0.00	0.00	0.00	0.00	0.00	0.00	0.00	0.00	0.00	1.42	28.8

Fig. 6.4 Monthly water volume contribution of PV and battery in Scenario #3.

The monthly pumping water contribution of the PV and battery of the proposed PVPS in Scenario #3 is shown in Fig. 6.4. The contribution of the battery is often zero in most months of the year, except January, November, and December. The maximum contribution for the battery occurs in January (around 29.9 m^3) while the minimum contribution is in November (around 1.42 m^3). It is clear that the contribution of the battery within a year is very little (0.42%) as compared to the PV array (99.58%).

Scenario #4 is similar to Scenario #3, but here the battery is replaced by a 1.5 kVA DG and an AC/DC converter with an efficiency equal to 90%. In this scenario, the priority is for the PV array and storage tank to provide the hourly water demand. The DG is turned on when the PV array and storage tank fail to provide the required amount of water per hour. The daily technical performance of the proposed Scenario #4 of the PVPS for a year is shown in Fig. 6.5. The first two subplots of Fig. 6.6 are similar to those in Scenario #3. Also, the daily Q_d and LLP are zero throughout the year, as illustrated in the fourth and sixth subplots of Fig. 6.5. The eighth and ninth subplots of Fig. 6.5, respectively, show the fuel consumption in liters (L) and the amount of CO_2 emissions in kg by considering that the DG produces around 2.5 kg/L of CO_2. The DG is used for only 8 days during the year. On Dec. 23, the system depends on the PV array and storage tank to meet the daily water demand, where the level of water in the storage tank on December 22 is 13.889 m^3. The average hourly pumping water volume

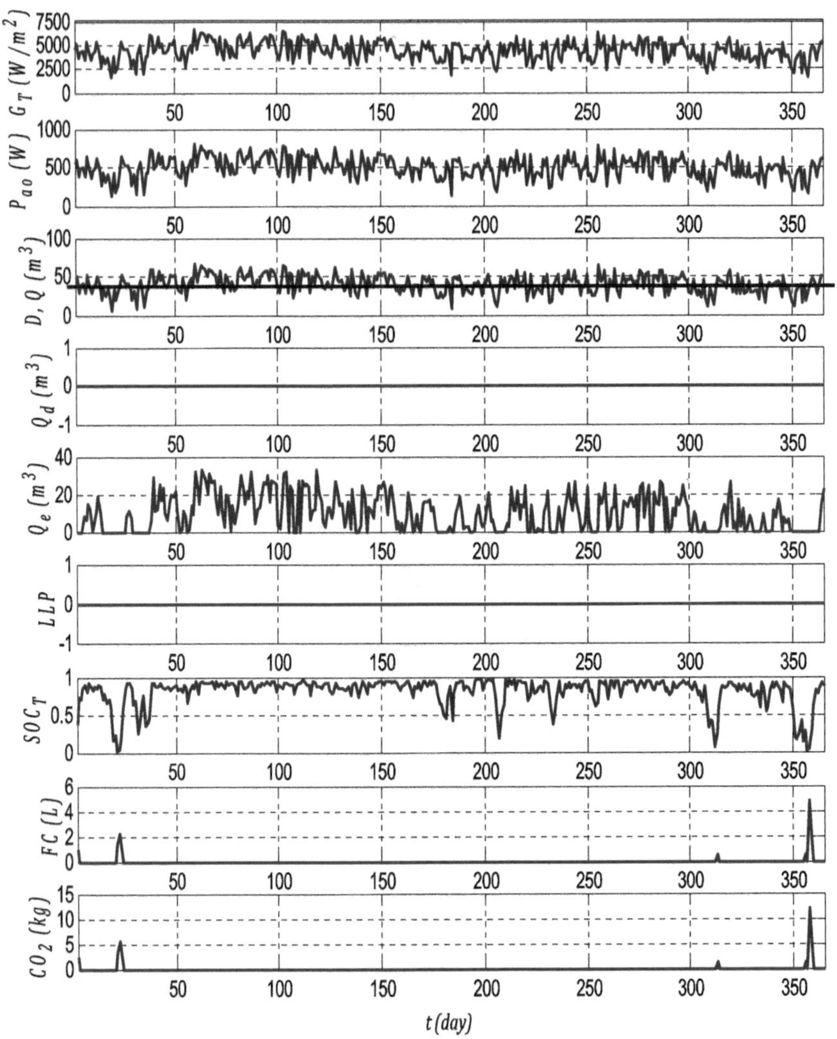

Fig. 6.5 Daily performance of PVPS of Scenario #4 for a year.

by Scenario #4 of the PVPS is 3.311 m^3, where the total fuel consumption and CO_2 emissions during the year are 12.539 L and 31.347 kg, respectively. According to Table 6.2, the minimum fuel consumption and CO_2 emissions, which occur on November 9 and December 22, are around 0.432 L/day and 1.081 kg/day, respectively. The maximum fuel consumption and CO_2 emissions occur on December 24, at around 4.756 L/day and 11.89 kg/day, respectively.

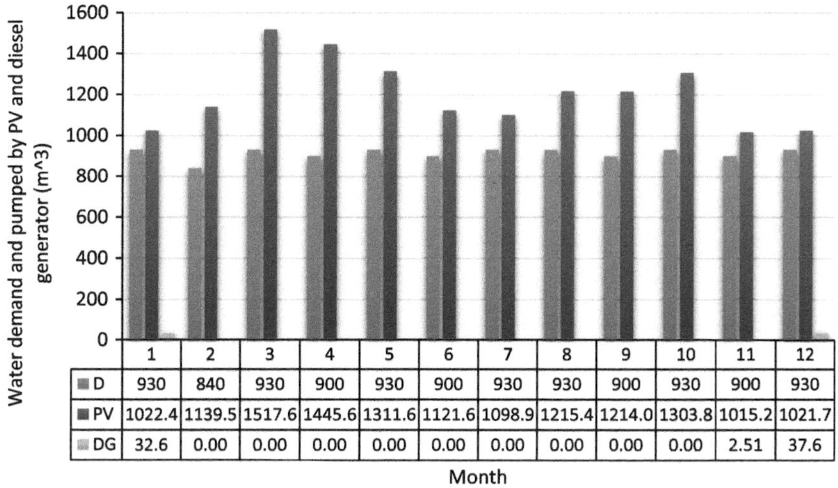

Fig. 6.6 Monthly pumping water volume contribution of PV and DG of PVPS of Scenario #4.

Fig. 6.6 shows the monthly contribution of the PV array and DG in the pumping water volume. For most months of the year, the DG contribution is zero, except in January, November and December with around 32.6, 2.51, and 37.6 m^3, respectively. This small DG contribution represents 0.5% of the productivity of the system for a year. It is clear that the DG contribution is greater than the battery contribution in Scenario #3 as compared to Scenario #4, where there are two operation modes which are standalone PV array mode and DG operation mode. In other words, the PV and DG did not work together at the same time in Scenario #4.

The proposed Scenario #5 of the PVPS is similar to Scenario #1, except that the storage tank is expanded to 80 m^3 to reduce the deficit days within the year. The system offers better performance with the *LLP* equal to zero most days of a year, except on January 1 as shown in Table 6.2. The proposed PVPS offers daily *LLP* and Q_d with values equal to 0.145 and 4.338 m^3, respectively, on Jan. 1, due to the lack of solar irradiance in the first two hours of the morning. The average hourly pumping water volume in Scenario #5 is still 3.297 m^3 because the configuration of the PV array is unchanged.

The configuration of Scenario #6 is similar to Scenario #5, except that the storage tank is initially full of water to overcome the deficit that occurs during the first hours of the morning of Jan. 1. The daily technical performance of the PVPS of Scenario #6 is shown in Fig. 6.7. It is clear that the

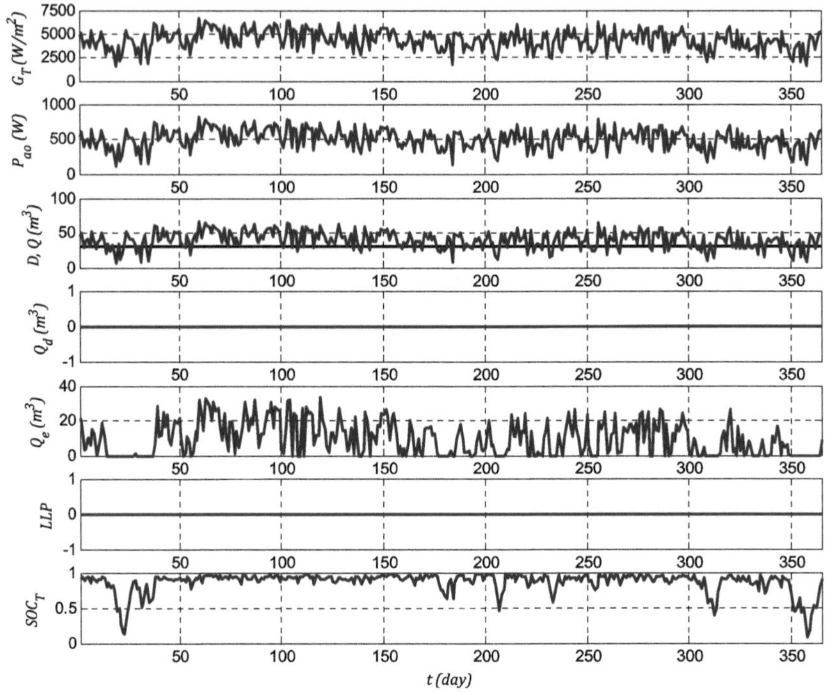

Fig. 6.7 Daily performance of PVPS of Scenario #6 for a year.

daily *LLP* and Q_d are zero during the year. This result is supported by Table 6.2, where the deficit on Jan.1 is overcame by using the initial full storage tank. According to Table 6.2, the amount of water in the storage tank at the end of the first operation day (January 1) becomes 73.029 m^3. The average hourly pumping water volume of the proposed PVPS in Scenario #6 is 3.297 m^3.

The last scenario is a pumping system powered only by a DG. A 30 m^3 empty storage tank is used to store the pumped water. The system pumps an hourly water volume of 2.505 m^3. The daily performance of the system for specific days is illustrated in Table 6.2. The system consumes 5.188 L/day of fuel to pump the required water for the load. Furthermore, the daily CO_2 emissions from the system are 12.971 kg. The state of charge of the storage tank becomes 0.736 at the end of the year.

6.4.2 Cost analysis

A detailed cost analysis of various proposed PVPS scenarios is presented in the following sections. The analysis is done based on the *LCC* method that is explained in Chapter 5. This analysis shows the cost contribution of each component of a system as well as the percentage of economic effect of the *LCC* elements of various scenarios. The cost, efficiency, and lifespan of various components are illustrated in Table 6.4. It is worth mentioning that the economic data of the PVPS used in Section 5.3, are also adopted in this chapter.

In Scenario #1, the cost analysis indicates that the civil and installation work costs constitute 40.36% of the *LCC* of the system, as shown in Fig. 6.8. The PV array life cycle cost is 27.55% and the DC/DC converter represents the lowest cost component with 6.8% of the *LCC* of the system. The initial, replacement & maintenance, and operation costs are 85.46%, 7.13%, and 7.41% of the *LCC* of the PVPS, respectively. It is clear that

Table 6.4 Specifications, cost, and lifetime of various components.

Item	Value
Battery:	
Cost (USD/kWh)	121.5
Efficiency (%)	85
Life span (year)	6
Maintenance in first year relative to initial cost (%)	3
DG:	
Cost (USD/kW)	550
Life span (h)	15,000
Engine oil cost (USD/L)	5.235
Oil filter cost (USD)	3
Air filter cost (USD)	3
Diesel filter cost (USD)	3
Spark plug cost (USD)	2
Fuel cost (USD/L)	0.39
Fuel inflation rate (%)	5
Engine oil lifetime (L/100 h)[a], (L/6 months)[b]	0.6
Oil filter lifetime (/100 h)[a], (/6 months)[b]	1
Air filter lifetime (/100 h)[a], (/3 months)[b]	1
Diesel filter lifetime (/100 h)[a], (/6 months)[b]	1
Spark plug lifetime (/100 h)[a], (/6 months)[b]	1
AC/DC converter and regulator:	
Cost (USD/W)	0.5
Life span (year)	10
Efficiency of AC-DC converter (%)	90
Efficiency of regulator (%)	95

[a]For DG standalone energy source.
[b]For DG and PV hybrid energy source.

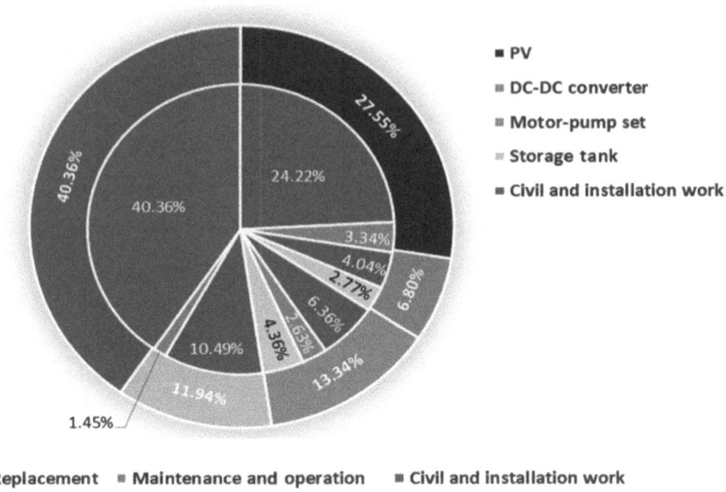

Fig. 6.8 *LCC* and cost contribution of various components of PVPS in Scenario #1.

Scenario #1 requires few replacement & maintenance and operation costs. On the other hand, the initial cost is high due to the PV array representing an expensive component in the system as well as the high cost of the civil and installation work.

The *LCC* elements and the cost contribution of various components of Scenario #2 are illustrated in Fig. 6.9. The civil and installation work cost of the PVPS is relatively high (37.61%), and the *LCC* of the PV array represents

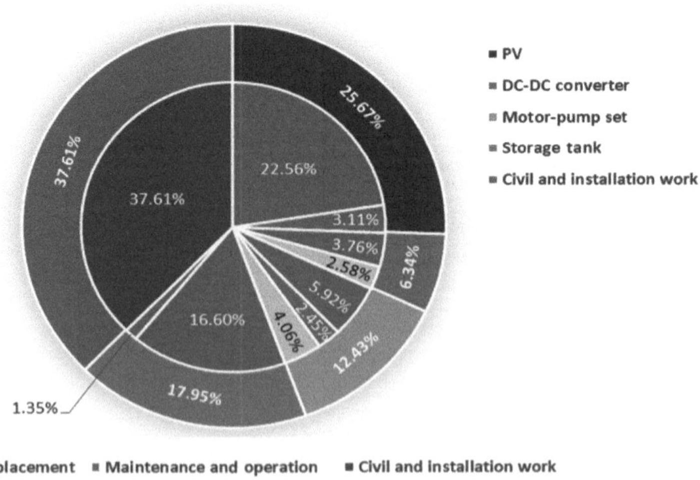

Fig. 6.9 *LCC* and cost contribution of various components of PVPS in Scenario #2.

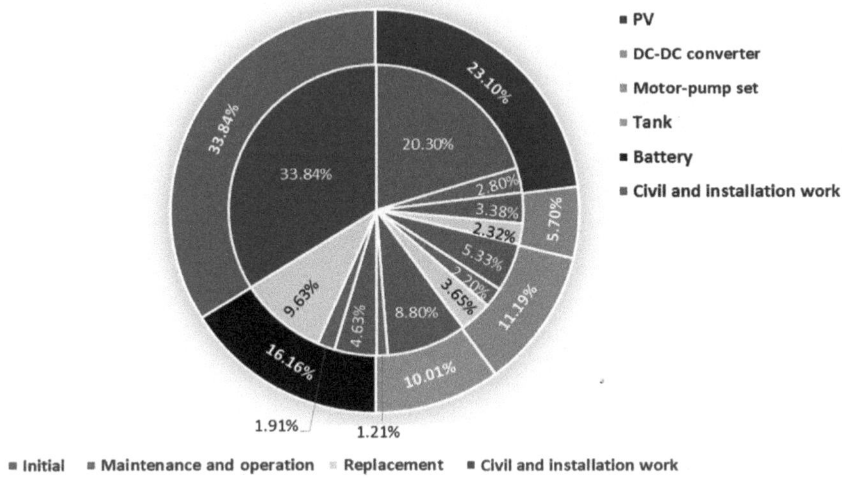

Fig. 6.10 *LCC* and cost contribution of various components of PVPS in Scenario #3.

25.67% of the total *LCC* of the system. The DC/DC converter is still the cheapest component in the PVPS with a 6.34% contribution to the *LCC* cost of the system. The initial capital, replacement & maintenance, and operation costs are 86.46%, 6.64%, and 6.9% of the *LCC*, respectively. It is worth mentioning that the initial cost of the PVPS of Scenario #2 is higher than that in Scenario #1, in spite of both scenarios having the same configuration. The increase in the initial cost of Scenario #2 is due to the cost of initially filling the storage tank with water.

Fig. 6.10 shows the main elements of *LCC* and the cost contribution of various components of the proposed PVPS in Scenario #3. The civil and installation work cost is the highest part of the *LCC* of the system (around 33.84%). Meanwhile, the *LCC* of both the PV array and the battery bank represents 23.1% and 16.16% of the *LCC* of the PVPS, respectively. The DC/DC converter constitutes 5.7% of the *LCC* of the system; it is the cheapest component in the PVPS of Scenario #3. According to Fig. 6.10, the total initial capital cost of the PVPS constitutes the main element of *LCC*, around 76.28%. The replacement & maintenance and operation costs constitute 15.6% and 8.12% of the *LCC* of PVPS of Scenario #3, respectively.

The *LCC* of each component of the PVPS proposed in Scenario #4 is represented in Fig. 6.11. In this scenario, the pumping system is powered by a PV array as well as a DG, where the DG is used on the deficit hours during a year. Therefore, the *LCC* of the DG is relatively low compared with that

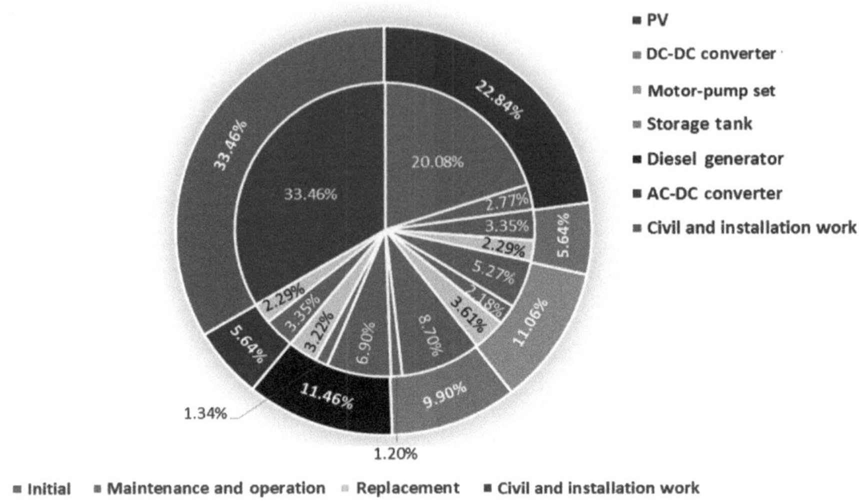

Fig. 6.11 *LCC* and cost contribution of various components of PVPS in Scenario #4.

of the pumping system powered by DG only. The *LCC* of the DG constitutes 11.46%. The *LCC* of the PV array and the civil and installation work costs are 22.84% and 33.46% of the total *LCC* of the system, respectively. The *LCC* of the converters represents the lowest cost compared to the other components of the system. According to Fig. 6.11, the initial, replacement & maintenance, and operation costs for the lifespan of the PVPS are 81.11%, 11.4 and 7.49%, respectively.

The cost contributions of various components of the PVPS of Scenario #5 are approximately similar to those values in Scenario #2. The *LCC* of the civil and installation work and the *LCC* of the PV array are the highest costs compared to the other components, with 37.92% and 25.89% of the total *LCC* of the system, respectively (see Fig. 6.12). The initial, replacement & maintenance, and operation costs slightly differ from those in Scenario #4, with 85.61%, 6.7% and 7.69%, respectively.

The civil and installation work for the *LCC* of the PV array and the storage tank are high compared to the other parts that constitute the *LCC* of the PVPS in Scenario #6, as indicated in Fig. 6.13. The DC/DC converter represents the cheapest component in Scenario #6, where *its LCC* does not exceed 5.97%. It is noticed that the *LCC* of the storage tank is increased because the size of the storage tank is increased. Also, the cost of initially filling the storage tank by water is added to the initial cost of the tank. The initial capital cost is still represented as the highest (around 86.55%)

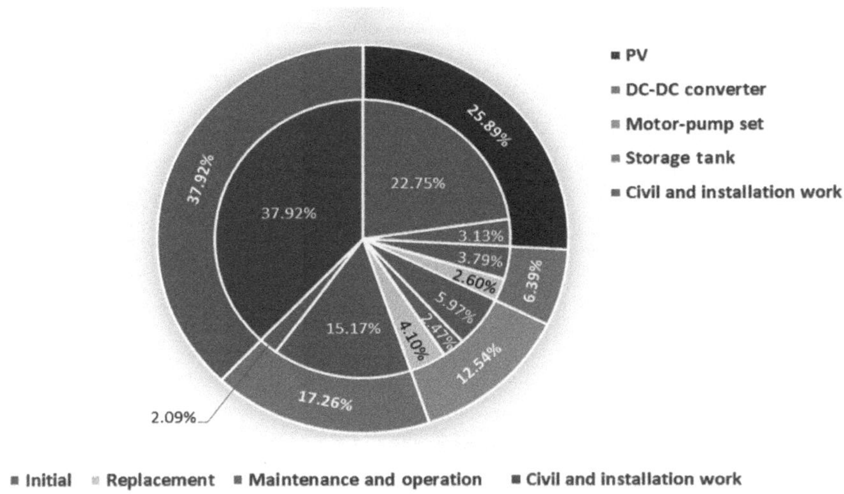

Fig. 6.12 *LCC* and cost contribution of various components of PVPS in Scenario #5.

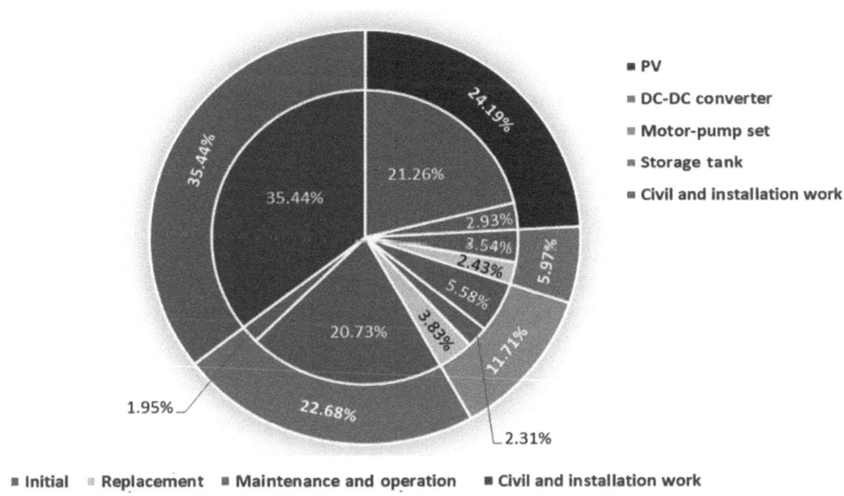

Fig. 6.13 *LCC* and cost contribution of various components of PVPS in Scenario #6.

between the other parts of the *LCC* (replacement cost around 6.26% and maintenance and operation cost around 7.19%).

The DG in this scenario represents the costly component in the proposed system due to the high maintenance and operation costs, which represents 77.69% of the total cost of the system as shown in Fig. 6.14. The AC/DC converter represents 2.29% of the total cost, and is the cheapest component

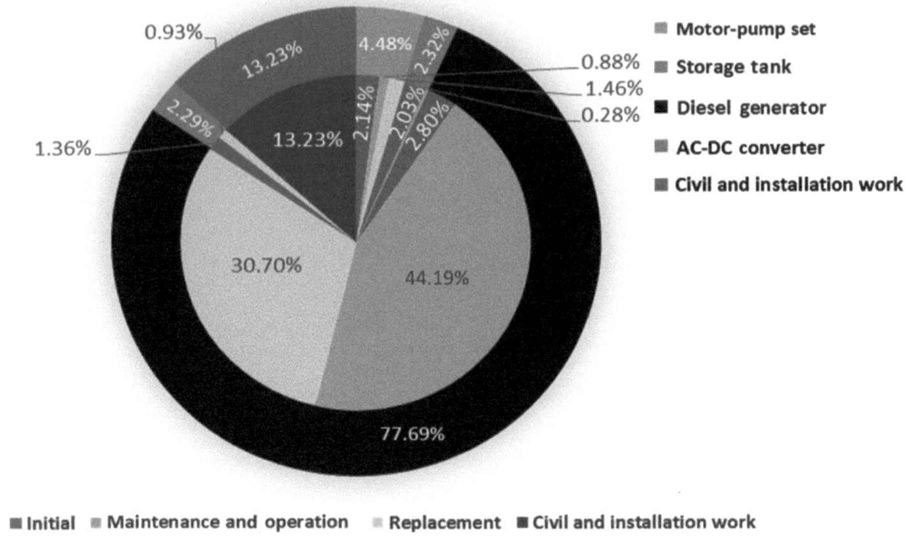

Fig. 6.14 *LCC* and cost contribution of various components of PVPS in Scenario #7.

in the system. The maintenance and operation costs are 45.36% of the total *LCC* of the system. Furthermore, the initial and replacement costs represent 21.55% and 33.09%, respectively.

6.4.3 Technoeconomic comparison

In this section, a technoeconomic comparison will be conducted between the various scenarios to identify the best scenario based on technical and economic aspects. All the proposed scenarios are run based on hourly meteorological data for a year. The productivity and *COU* of various scenarios are illustrated in Table 6.5. The third and fourth scenarios offer yearly *LLP* and Q_d that are zero, but the cost of the unit is higher than 0.053 USD/m^3 for both scenarios; also, the fourth scenario emits at least 31 kg of pollutant. The last scenario (Scenario #7) also offers good technical performance with yearly *LLP* and Q_d that are zero, but the *COU* is higher than other systems by eight cents. Moreover, the system pollutes the environment with 4.734 ton of CO_2. In addition, the pumping systems based on DG are not reliable because of their frequent faults. The PVPSs of Scenarios 1, 2, and 5 offer an *LLP* greater than zero and a *COU* range from [0.045, 0.049] USD/m^3. The best PVPS is that proposed in Scenario #6, as it offers better technical performance where the *LLP* and Q_d are zero for a year with

Table 6.5 Productivity and *LLP* of various scenarios for a year.

Scenario	*LLP*	*COU* (USD/m^3)	Q (m^3)	Q_d (m^3)	Q_e (m^3)	FC (L)	CO_2 (kg)
1	0.00502	0.04530	14,439.173	55.0150	3500.0837	0.00000	0.00000
2	0.00463	0.04883	14,439.173	50.6775	3547.7462	0.00000	0.00000
3	0.00000	0.05376	14,417.454	0.00000	3427.6976	0.00000	0.00000
4	0.00000	0.05437	14,498.272	0.00000	3505.2048	12.53865	31.34663
5	0.00040	0.04786	14,439.173	4.33755	3421.4063	0.00000	0.00000
6	0.00000	0.05158	14,439.173	0.00000	3497.0687	0.00000	0.00000
7	0.00000	0.13437	10,967.067	0.00000	0.00000	1893.7689	4734.4222

reasonable *COU* and excess water volume. Furthermore, Scenario #6 is environmentally friendly compared to the systems in Scenario 3, 4, and 7.

6.4.4 Sensitivity of *COU*

The effect of some parameter variations on the cost of the system were studied, analyzed, and evaluated in this section. These parameters are the initial cost of the PV array, the storage tank, the battery, the DG, and the fuel cost. The parameter variations were considered from 50% to 150%, where the 100% case represents the base and the parameter values of this case are given in Table 6.4. The analyses are conducted for Scenarios 6, 3, and 4 only because they are, respectively, the best scenarios from a technical aspect among the seven proposed scenarios. The effects of variations of the initial capital costs of the PV array and storage tank were considered to evaluate their effect on the *COU* of the system in Scenario #6. It is clear from Fig. 6.15 that the *COU* in this system is more sensitive to the initial capital cost of the PV array than that of the storage tank.

Fig. 6.16 shows the effect of the initial capital costs of the PV array, storage tank, and battery on the *COU* of the proposed system in Scenario #3. According to Fig. 6.16, the *COU* is also more affected by changing the initial capital cost of the PV array. The *COU* is less sensitive to the variation of the

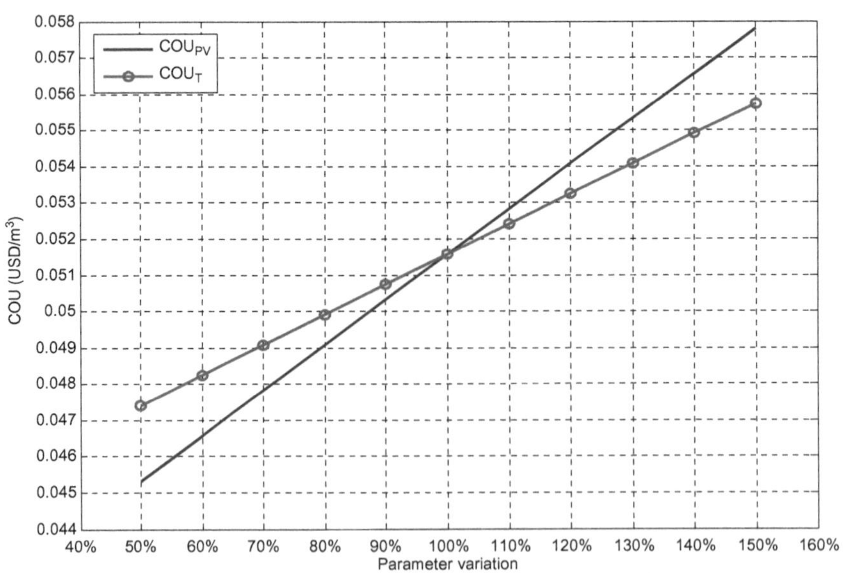

Fig. 6.15 Sensitivity of *COU* to variation of various parameters in Scenario #6.

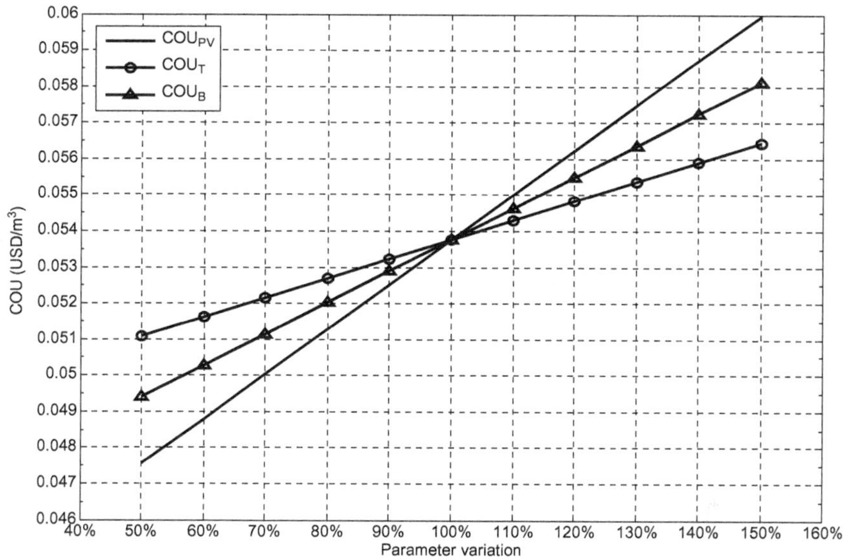

Fig. 6.16 Sensitivity of *COU* to variation of various parameters in Scenario #3.

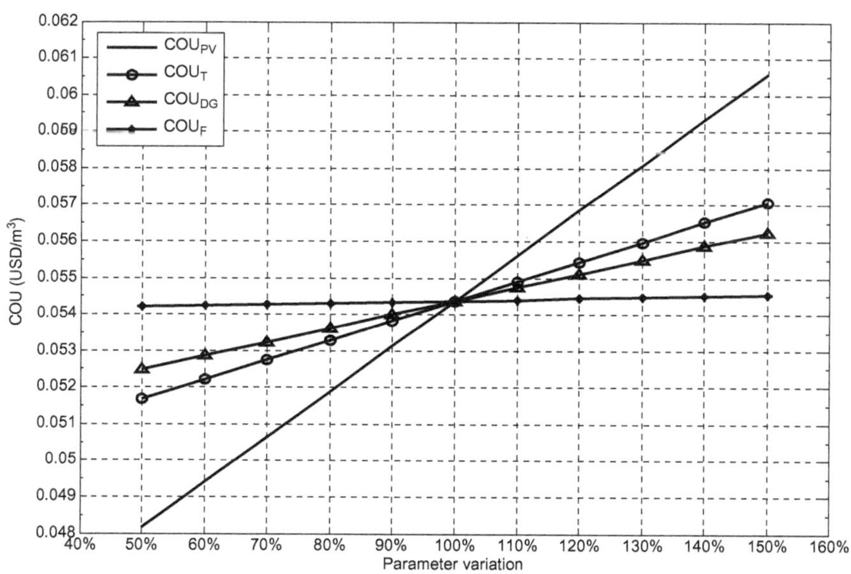

Fig. 6.17 Sensitivity of *COU* to variation of various parameters in Scenario #4.

initial capital cost of the storage tank. This is due to the initial cost of the storage tank in the base case being relatively low as compared to the initial cost of the PV array.

The impact of the initial costs of the PV array, storage tank, DG, and the fuel price on the *COU* are shown in Fig. 6.17. The initial capital cost of the PV array has a higher impact on the *COU* while the *COU* is more sensitive to the initial cost of the water storage tank than that of the DG. The effect of variation of fuel price on the *COU* is very little as compared to other parameters. This is due to the fact that the contributions (operation hours) of the DG in this scenario are very little (around 29 h/year) as compared to the PV array.

CHAPTER 7

Photovoltaic pumping system for water treatment of artificial bonds in agriculture fields

Contents

7.1 Introduction

In general, there are many agriculture zones that suffer from limited water sources. Farmers, therefore, use another solution to collect and reserve water, where they dig to make artificial ponds to collect water from rainfall, nearby artesian wells, and any available excess water [1, 2]. An artificial pond is a very good collecting technique, as it is an integrated way to collect and store runoff surface water from rainfall that runs on roofs of greenhouses. Agricultural plastic greenhouses are one of the facilities that can be used to collect rainwater. Each agricultural greenhouse with an area of 1 acre can collect about 300 m^3 annually of fresh rainwater, which is suitable for planting all kinds of crops. However, artificial ponds are usually not preferred and sometimes are prevented for environmental reasons because these ponds create many environmental problems. Artificial ponds cause a bad view, a stinky smell, and collect many types of insects, flies, frogs, cockroaches, and worms. The most dangerous are the

https://doi.org/10.1016/B978-0-12-821231-8.00007-X

insects and microorganisms that are attracted by the algae formed on the surface. These insects and microorganisms may carry viral and bacterial diseases that destroy crops and affect their quality [3, 4].

Here also, farmers usually try to solve the growing algae problem by adding many types of fish such as tilapia, lionhead, and koi that feed on algae and some insects that the artificial pond attracts. However, this solution is not suitable because it affects fish badly and can kill them due to oxygen depletion and water pollution [5, 6]. Based on that, there is a dire need to treat the water in these ponds so as to address this environmental issue.

In general, algae pollution has become a global issue recently. The solution is to treat and control the growth of algae by such methods as biological, physical, and chemical controls. Zhang et al. [7] studied the ultrasonic removal of *Microcystis aeruginosa*. It was chosen because it is a major bloom-forming and poisonous algae species that is widely found in natural waters. Similarly, Heng et al. [8] reported the effectiveness of ultrasonic irradiation on algae removal by coagulation. In addition, Wang et al. [9] studied removing algae from lake water and the attendant water quality changes using ultrasound under different conditions. Moreover, Zhang et al. [10] introduced a new technology for the effective removal of algae cells from the source water in a water treatment based on the sonication-coagulation method. On the other hand, Ghernaout et al. [11] studied algae and cyanotoxin removal by coagulation-flocculation. Hu et al. [12] studied the feasibility of using coagulation to treat the slightly polluted algae-containing raw water of the Pearl River; this combined ozone preoxidation with poly aluminum chloride (PAC). Hoko and Makado [13] presented the sampling and analysis of parameters according to APHA standards. It is claimed in this research that algal removal is increased with increasing contact times. Meanwhile, Shen et al. [14] studied algae removal by drinking water treatment of chlorination coupled with coagulation. Phoochinda and Whit [15] investigated algae removal using froth flotation. It was reported that decreasing the pH values of the algal suspension may increase the algal removal efficiency up to 80%. In another work, Kwon et al. [16] studied removing algae and turbidity by floating-media and sand filtration. Finally, Shehata et al. [17] claimed that the algal removal rate increased when using oxidants combined with alum-potassium permanganate, as this is the most effective treatment combination for algal removal.

Based on the reviewed research, there are several possible approaches to control algae. Management practices for nuisance algae are divided into two major categories: nutrient manipulation and direct control techniques.

Nutrient manipulation is considered the best approach for the long-term control of the algal problem. There are situations for which significant nutrient reduction is impractical or ineffective. Under these conditions, direct control of the algal biomass may be the only alternative available.

Nutrient manipulation can be done, for example, by mechanical mixing (water circulation) using surface-mounted pumps. This mixing of the water column disrupts the behavior of cyanobacteria to migrate vertically in addition to limiting the accessibility of nutrients. The disadvantage of the mixing or circulation of water is often the high maintenance required for the system and the efficiency of the system, which depends on water quality. Moreover, if the circulation system runs for 12 h per day, the total energy required is 0.525 kWh/m^2.yr. Thus, all these methods require high levels of water pumping, pond mixing, or aeration, which implies high energy consumption. Meanwhile, removing algae from lake water using ultrasound under different conditions could be efficient. However, to reach a removal efficiency higher than 95%, high frequencies with relatively large power sources (20–80 W) are required. In addition to that, the major disadvantage of UV disinfection is the UV intensity, which decreases sharply with its passage in water and its decrease is even more significant with high water turbidity.

On the other hand, direct control methods can be done using chemicals by the coagulation method. It was shown that this method requires lower operating costs and little maintenance in terms of biological operation. The coagulant is rapidly and thoroughly dispersed on dosing by adding it at a point of high turbulence. The advantages of coagulation are that it reduces the time required to settle suspended solids and is very effective in removing fine particles that are otherwise very difficult to remove.

Solar water pumping is based on photovoltaic (PV) technology that converts solar energy into electrical energy to run a DC or AC motor-based pump. The use of solar photovoltaic energy in pumping is considered to be a primary resource, where direct solar radiation may reach 1000 W/m^2. The advantages of PV-powered pumps are low maintenance, no pollution, easy installation, reliability, the possibility of unattended operation, and the capability to be matched with solar radiation because, in most cases, the water demand increases during summer when solar radiation is at its maximum. Meanwhile, the disadvantages are the high initial cost and variable water production. After all, many researchers have concluded the superiority of a PV-based pumping system as compared to other choices such as a diesel generator in remote areas [17]. Moreover, the performance of PV-based pumping systems for agricultural applications was also shown by other researchers.

Finally, the feasibility of a PV pumping system was investigated and reported positively in some studies.

Based on that, in this chapter an efficient method for irrigation water treatment based on a photovoltaic pumping system for artificial ponds in the Jordan Valley is proposed. The method is presented by a small PV pump-driven system that pumps chemical material into the water of an artificial pond. The chemical material is pumped with high pressure to rotary nozzles that are fixed on the pipe, which is installed inside the pond at a specific depth so as to ensure the distribution of the material to all corners of the pond.

7.2 Proposed chemical material for water treatment

In this research, Tammun, a village in the Tubas governorate in Palestine, was adopted as a case study. There are between 300 and 400 artificial ponds in the Tammun village valley with different sizes; they are in the range of 10–15 m length, 4–8 m width, and 1.5–2 m depth.

In general, the farmers in this village create artificial ponds to address the problem of water shortages. During summer, water in these artificial ponds is collected from five artesian wells around the village while in winter, water comes from rainfall. It is worth mentioning that the rainfall in Palestine usually starts in the middle of October and continues to the end of April. The amount of rainfall in the Jordan Valley and Dead Sea area has an average of 100 mm as a minimum value.

7.2.1 Sample collection and analysis

A specific site was selected in this village and samples were collected from the site with a storing temperature of 2–8°C, as shown in Fig. 7.1.

After that, sample analysis was done in the analysis and calibration unit at An-Najah National University, as shown in Table 7.1. According to the test, the percentages of pH, SO4, PO_4, Cl, NO_3, Ca, Na, $CaCO_3$, K, TDS, and Mg were within the normal range except for chemical oxygen demand (COD), which was 7200 mg/L. This value, in fact, is very high, as the allowable value is a maximum of 250 mg/L.

COD is the amount of oxygen that is consumed in the oxidation of organic matter. It is used to measure the total amount of organics formed due to a high concentration of phosphorus in the pond. High concentrations of phosphorus may result from poor agricultural practices, runoff from urban areas and lawns, leaking septic systems, or discharges from sewage treatment

Fig. 7.1 Capture of the site.

Table 7.1 Sample testing results.

Test	Units	Results	Limits	Ref
pH	–	6.76		SMWW[a]
Nitrate (NO_3)	mg/L	1.3		SMWW[a]
Sulfate (SO_4)	mg/L	30.1		SMWW[a]
Phosphate (PO_4)	mg/L	0.8		SMWW[a]
Chloride (Cl)	mg/L	100.0		SMWW[a]
Total alkalinity (as $CaCO_3$)	mg/L	240		SMWW[a]
Total hardness (as $CaCO_3$)	mg/L	290		SMWW[a]
Total dissolved solid (TDS)	mg/L	454.5		SMWW[a]
Calcium (Ca)	mg/L	56		SMWW[a]
Magnesium (Mg)	mg/L	36.5		SMWW[a]
Sodium (Na)	mg/L	36		SMWW[a]
Potassium (K)	mg/L	27		SMWW[a]
Chemical oxygen demand (COD)	mg/L	7200		SMWW[a]

[a]SMWW, standard method of water and waste water.

plants. Phosphorus is considered a vital nutrient for converting sunlight into usable energy, and is essential to cellular growth and reproduction causing algae blooms.

7.2.2 Proposed treating material

This research aimed at conducting nutrient manipulation to kill algae. According to many studies, phosphorus is the main reason for algae growth. Thus, the aim was to develop a material that dissolves phosphorus so as to get rid of algae. Many studies show that the most used chemicals for controlling phosphorus in ponds are lime and alum. Lime is used for the treatment of raw wastewaters and could lead to an 80–90% removal of phosphorus. The use of lime as a coagulant appeared to be more efficient. As for alum, it is concluded that the use of alum as a coagulant is moderately effective in algal removal. Alum includes many components, such as potassium alum sulfate, chosen in this paper because aluminum sulfate is a coagulant and potassium is a very important fertilizer for plants. To do so, potassium aluminum sulfate is utilized. Potassium aluminum sulfate is a chemical (potash alum) with a chemical formula of $KAl\ (SO_4)_2.12H_2O$. This substance was particularly chosen because it gets rid of phosphorus, which is the main reason for algae growth. This material was obtained after several laboratory experiments, including using other materials that are common in the purification of algae such as lime.

7.3 Materials and methods

Solar water pumping systems may contain DC and AC motors. DC motors are directly connected to the PV generator. According to the water pumping requirements, solar water pumps may be divided into three categories: submersible, surface, and floating [18]. PV power must cover the power demand of the pump adequately. This is determined by the relationship between the required discharge flow, the total head, and the pump's efficiency. Positive displacement pumps are preferred for large pumping heads, whereas centrifugal pumps are most commonly used for low pumping heads. In this article, a DC surface centrifugal water pump is used [18].

The first step in designing the proposed system is tank sizing as well as nozzle and pipe material selection. Meanwhile, the second step is to calculate the flow rate and the total dynamic head. After that, the required power is calculated. Finally, the PV generator capacity and system control strategy are decided.

7.3.1 Materials selection

Many factors and constraints need to be considered in selecting the materials for the proposed system. These constrains are the chemical material nature, the ease of use, and the economic aspect. The tank, pipes, rotary nozzle, and fittings material are PVC plastic because it is available and does not react with other materials such as metal and steel. Moreover, it is lighter, easier to shake in the tank, and cheaper than metal and steel.

The required number of rotary nozzles is estimated to be three nozzles because one of them can cover a distance of 5 m with a radius of 360 degrees and a working pressure of 2.4 bar. It is suitable to overcome the total pressure exerted against the inlet area 0.5″ of the nozzles, which is almost equal to 1.8 bar.

7.3.2 Flow rate estimation

The flow rate is calculated using the following equation:

$$Q = \frac{V}{t} \tag{7.1}$$

where Q is the flowrate, V is the volume of the supplying tank of chemical substance, and t is the time to evacuate the tank.

To estimate the diameter of the pipe, the velocity (v) is set to be in the recommended range for centrifugal pump suction and discharge velocities, which are in the range of 0.6–1.2 m/s and 2.1–7.6 m/s, respectively. This assumption is applied for both the suction and discharge range. The pipe chosen is PVC (40 schedule) and its cross-sectional area is calculated using the following equation:

$$Q = VA \tag{7.2}$$

where Q is the flowrate, v is the velocity in the pipe, and A is the cross-sectional area for the pipe. The results above match the centrifugal pumps rules below:

$$D_{\text{suction.}} > D_{\text{Discharge}} \tag{7.3}$$

$$V_{\text{suction.}} < V_{\text{Discharge}} \tag{7.4}$$

7.3.3 Total dynamic head estimation

The total dynamic head (TDH) is the total equivalent vertical distance that the pump must move the water, or the pressure that the pump must overcome to move the water to a certain distance, as shown in Fig. 7.2.

The losses in pipes are calculated as follows:

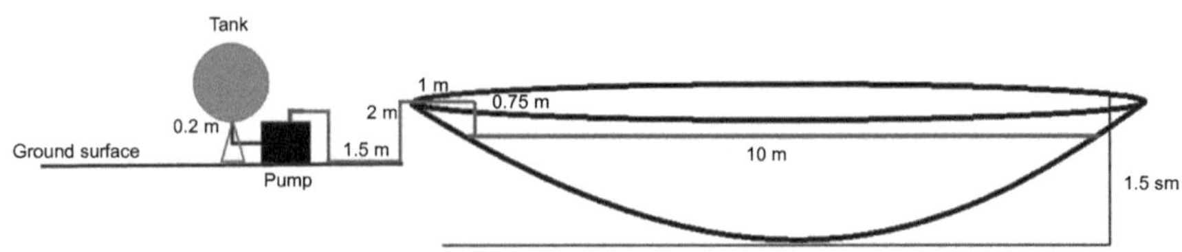

Fig. 7.2 The typical head of the water pump.

$$R_e = \frac{vD}{U} \tag{7.5}$$

where R_e is the Reynolds number, v is the velocity in the pipe, D is the diameter of the pipe, and U is the kinematic viscosity, which is 1.31×10^6 m^2/s for the chemical solution. The friction coefficient (f) can be found using a modified equation:

$$f = \frac{0.25}{\left[\log\left(\frac{k}{3.7D} + \frac{5.74}{Re^{0.9}}\right)\right]^2} \tag{7.6}$$

where k is the roughness coefficient for PVC pipes, which is 0.0015 mm; D the is pipe diameter; Re is the Reynolds number; and L is the length of the pipes.

To calculate dynamic head losses in the pipes DH, Eqs. (7.7) and (7.8) are used:

$$\mathrm{DH} = \frac{K v^2}{2g} \tag{7.7}$$

where K is the resistance coefficient, v is the velocity in the pipe, and g is the acceleration due to gravity. To calculate K for the pipes, Eq. (7.8) is used:

$$K = \frac{fL}{D} \tag{7.8}$$

where K is the resistance coefficient, f is the friction coefficient, L is the pipe length, and D is the diameter of the pipe. The K values for suction and discharge are 0.405 and 67.15, respectively. As for the pipe, the DH is calculated for suction and discharge, and it equals 7.44×10^{-3} m and 15.11 m, respectively.

The losses in fittings K_{fittings} and DH have been calculated as well. K_{fittings} is the total resistance coefficient for fittings and it equals 13 because it equals 1 for a nonreturn valve (one item), 2 for a tee branch flow threaded (three items), 1.5 for a 90-degree bend threaded (seven items), and 6 for a water meter turbine wheel. The total dynamic head (HD) for plastic PVC fittings is 2.92 m and 0.11 m through a water meter. Finally, Bernoulli's equation is used to get the value of TDH as below:

$$\frac{P1}{\gamma}+\frac{v^{12}}{2g}+Z1-hl+ha=\frac{P2}{\gamma}+\frac{v^{22}}{2g}+Z2 \tag{7.9}$$

where P is pressure; $P_{1\text{-tank}}$ equals 13.005 Psi (89.7 kPa); $P_{2\text{-nozzle}}$ equals 26.175 Psi (180.5 kPa); γ is the specific weight of the material, which equals ρg where ρ is the potash alum solution density (1725 kg/m^3); ha is the head loss; v is the velocity of the fluid at the inlet/outlet; z is the elevation of the point above a reference plane that is specified; and g stands for the magnitude of the acceleration due to gravity = 9.81 m/s^2.

7.3.4 Pump selection and PV sizing

There are several types of electrical motors that can be utilized to run the pump, such as AC and DC. DC motors are an attractive option because of their compatibility with the power source and because their efficiency is usually higher than that of AC motors. Therefore, a DC motor is used for this application. The size of the water pump is calculated based on the following equation:

$$P=\frac{QHg\rho}{(\eta)} \tag{7.10}$$

where P is the pump's power, Q is the flowrate, H is the total dynamic head, ρ is the potash alum solution density (1725 kg/m^3), g is acceleration gravity (9.81 m/s^2), and η is the pump efficiency (Fig. 7.3).

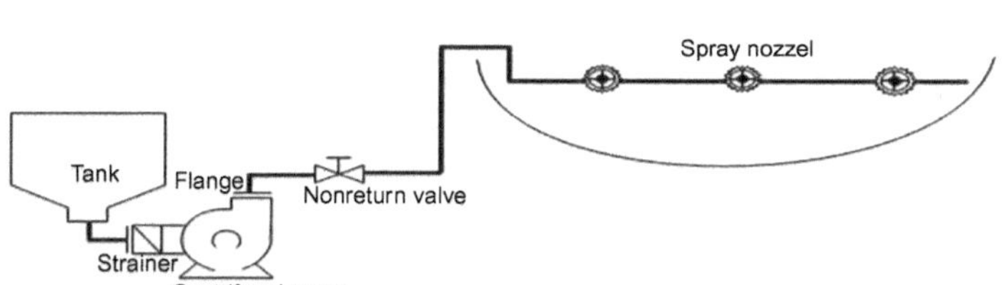

Fig. 7.3 Mechanical diagram of the system.

7.4 System effectiveness and performance

The testing of the proposed system was done at a lab temperature of 25°C using 200 mL and 100 mL beaker containers for mixing, a dropper, a scale to weigh the chemical material, safety goggles, gloves, and a lab coat.

In general, lime is recommended by some studies. However, using lime alone decreased the pH value where it reached 12.3 while the normal pH value for irrigation water ranges from 6.5 to 8.4. To solve this problem, potash alum was added to adjust the pH value within the required limit. Using potash alum alone gave good results as well as an acceptable pH value of 7.45, and solved the problem of COD. The chemical material is added and then a rapid mixing is applied to dissolve the chemical and distribute it evenly throughout the water. This practice aims to bind the phosphorous with the potash alum and turn it to sediment in the bottom of the water with a settling time of 15 min. The purified water that contains the sediment should not be used in irrigation until the water is passed through a filter. As a result, the use of the proposed material decreases the COD from 7200 to 200 mg/L. Such a removal ratio is considered high as compared to other studies. The removal ratio of algae, for example, by using potassium ferrate (K_2FeO_4) is 92.5% with the contact time set to 1 min. Meanwhile, the potassium permanganate ($KMnO_4$) removal ratio is found to be 74.6% after 10 min of contact.

Anyway, jar tests showed that a potash alum amount of 14 g that is rapidly mixed with 100 mL of clean water is the recommended mixture ratio for water treatment. In this research, 200 mL of an untreated pond water sample are mixed with 0.5 mL of the recommended mixture with a concentration of 0.295 mol/L. This practice decreased the COD value from 7200 mg/L to around 95 mg/L, as mentioned earlier.

In this research, the adapted case study is an artificial pond with a volume of 120,000 L. Thus, the dosing volume material needed is 300 L of the recommended mixture. This is in order to achieve the goal for both algae removal and prevention of algae blooming. A tank volume of 20 L is considered because there are commercial tanks available with this size (water cooler bottles). The choice of a 20 L bottle will make it easy to carry and to shake.

The pumping time is assumed to be during the sun's peak (11 a.m.–2 p.m.) and for 3 h. Thus, in order to finish the 300 L dose in that time, the pumped time is set to be 4 min. Based on that, the flow rate is estimated using Eq. (7.1) to be $0.083 \times 10^{-3}\ m^3/s$.

The type of pipe chosen is PVC (40 schedules), which can handle 10 bar with a nominal pipe size for suction and discharge with cross-section areas equal to 0.75″ and 0.5″, respectively. According to Eq. (7.2), the velocity of the suction and the discharge process are set to 0.6 m/s and 2.1 m/s respectively. The Reynolds numbers for suction and discharge are found to be 6077.86 and 11,381.6, respectively, using Eq. (7.5). The friction coefficient (f) for suction and discharge is found to be 0.03585 and 0.03027, respectively, by using Eq. (7.6). The total dynamic head for pipes is 15.11744 m. The head loss was calculated from Eq. (7.9) and was found to be about 20 m.

All pumping is assumed to be done using an electric centrifugal DC pump with an overall power calculated to be 36.11 W, according to Eq. (7.10), with an efficiency of 77%. Because energy loss from the solar panels is in the range of 30–35%, a 50 W photovoltaic foldable solar panel is chosen.

Fig. 7.4A–C below show the purified water before treatment, after treatment, and the dead algae precipitated. It also shows the situation after getting rid of all algae and after filtering the treated water. Finally Fig. 7.5 shows a 3D

Fig. 7.4 The purified water result (A) Before treatment. (B) After treatment and the dead algae that was precipitated. (C) After getting rid of all algae participated in the bottom and the water became pure.

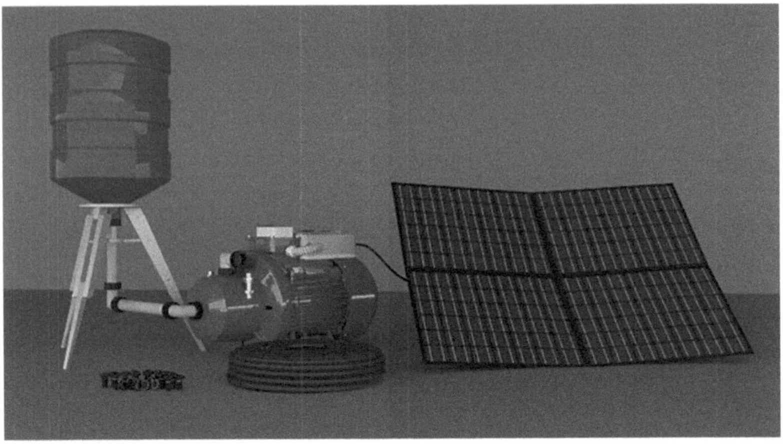

Fig. 7.5 Proposed design of the product.

Table 7.2 Proposed device cost.

Estimated parts	Required parts/ unit	Cost/ parts
Centrifugal pump	1	$50.00
Plastic tank (20 L)	1	$4.96
Pex pipes (16 mm, PE)	18 m	$4.96
90-degree elbows (elbow joiner)	7	$1.65
T-elbow	3	$1.65
Nonreturn valve	1	$3.31
Nozzles (rotor sprinkler)	3	$9.92
Chemical material ($KAl(SO_4)_2$)/bottle	2.8 kg	$32.40
Tank stand	1	$5.56
End line	1	$0.28
Screw	4	$1.10
Arduino, relay, and switch	1	$16.53
Sensor flow meter	1	$13.77
Clipper	1	$0.28
CB, DC surge arrestor and fuse	1	$13.77
Joints at inlet and outlet of the pump (coupling)	2	$1.10
Foldable 50 W solar panel	1	$50.00
Total price = $213		

design of the proposed product. As for the cost of the system, Table 7.2 shows a price estimation per unit of the proposed system. According to the table, the cost of the unit is about $213, which is considered affordable.

According to this price, the product is found to be affordable and cost effective, considering the product price and chemical material price.

References

[1] Muhsen DH, Khatib T, Nagi F. A review of photovoltaic water pumping system designing methods, control strategies and field performance. Renew Sustain Energy Rev 2017;68(P1):70–86.
[2] Gopal C, Mohanraj M, Chandramohan P, Chandrasekar P. Renewable energy source water pumping systems—a literature review. Renew Sustain Energy Rev 2013;25:351–70.
[3] Scott MJ, Gould F, Lorenzen M, Grubbs N, Edwards O, O'Brochta D. Agricultural production: assessment of the potential use of Cas9-mediated gene drive systems for agricultural pest control. J Responsible Innov 2018;5(sup1):S98–S120.
[4] Raheja AK. Assessment of losses caused by insect pests to cowpeas in northern Nigeria. PANS 1976;22(2):229–33.
[5] Nanninga H, Tyrrell T. Importance of light for the formation of algal blooms by Emiliania huxleyi. Mar Ecol Prog Ser 1996;136:195–203.
[6] Thronson A, Quigg A. Fifty-five years of fish kills in coastal Texas. Estuar Coasts 2008;31(4):802–13.
[7] Zhang G, Zhang P, Wang B, Liu H. Ultrasonic frequency effects on the removal of Microcystis aeruginosa. Ultrason Sonochem 2006;13(5):446–50.
[8] Heng L, Jun N, Wen-jie H, Guibai L. Algae removal by ultrasonic irradiation–coagulation. Desalination 2009;239(1–3):191–7.
[9] Li J, Long H, Song C, Wu W, Yeabah TO, Qiu Y. Study on the removal of algae from lake water and its attendant water quality changes using ultrasound. Desalin Water Treat 2014;52(25–27):4762–71.
[10] Zhang G, Wang B, Zhang P, Wang L, Wang AH. Removal of algae by sonication-coagulation. J Environ Sci Heal Part A 2006;41(7):1379–90.
[11] Ghernaout B, Ghernaout D, Saiba A. Algae and cyanotoxins removal by coagulation/flocculation: a review. Desalin Water Treat 2010;20(1–3):133–43.
[12] Hu WC, Wu CD, Jia AY, Chen F. Enhanced coagulation for treating slightly polluted algae-containing raw water of the Pearl River combining ozone pre-oxidation with polyaluminum chloride (PAC). Desalin Water Treat 2015;56(6):1698–703.
[13] Hoko Z, Makado PK. Optimization of algal removal process at Morton Jaffray water works, Harare, Zimbabwe. Phys Chem Earth, Parts A/B/C 2011;36(14–15):1141–50.
[14] Shen Q, Zhu J, Cheng L, Zhang J, Zhang Z, Xu X. Enhanced algae removal by drinking water treatment of chlorination coupled with coagulation. Desalination 2011;271(1–3):236–40.
[15] Phoochinda W, White DA. Removal of algae using froth flotation. Environ Technol 2003;24(1):87–96.
[16] Kwon D-Y, Kwon J-H, Jo G-J. Removal of algae and turbidity by floating-media and sand filtration. Desalin Water Treat 2014;52(4–6):1007–13.
[17] Shehata S, Badr S, Wahba S. Drinking water treatment options for eliminating freshwater algae. Int J Environ Stud 2002;59(6):679–88.
[18] Raghuwanshi SS, Khare V. Sizing and modelling of stand-alone photovoltaic water pumping system for irrigation. Energy Environ 2018;29(4):473–91.

CHAPTER 8

Environmental and social impact assessment methodology of megascale photovoltaic water pumping systems

Contents

Photovoltaic Water Pumping Systems
https://doi.org/10.1016/B978-0-12-821231-8.00008-1

8.1 Introduction

In any country, there should be a number of national policies, laws, and instruments that are available to support environmental management and the environmental impact assessment process. Environmental law, environmental assessment policy, and other sectoral sections in other legislation are the key instruments that cover environmental management in development sectors. The environmental assessment policy usually prescribes the process, procedures, and practices for conducting an environmental assessment and preparing environmental assessment reports.

Besides national environmental laws, environmental assessment policy should include industrial standards and mechanisms for the monitoring and enforcement of environmental regulations. However, in case of any missing issue at the national level, international environmental and social framework (ESF) and standards gap analysis should carried out.

For any project, there should be an implementing agency, which in this chapter is called the OWNER. Meanwhile, projects are usually tendered and awarded to a contractor/supplier for implementation. These contracts are the responsibility of the implementing agency. However, other stakeholders should have a technical role in the bidding process, including technical supervision during implementation. All bidding documents should include environmental and social management measures and all contracts will include subproject specific management plans to be implemented by contractors and subcontractors. Here, the implementing agency should have an environmental and social officer (ESO) who is responsible for environmental and social management.

Based on project activities and the analysis and evaluation of the environmental and social risks of the project, the environmental and social impact assessment (ESIA): (i) defines the screening process to be followed to determine the gaps to be addressed under the environmental and social instruments; (ii) identifies potential environmental and social risks and impacts; (ii) determines the appropriate environmental and social risk classification for each component; and (iii) identifies, mitigates, and monitors indicator measures.

8.2 Environmental and social risk screening

The potential negative impacts during PVPS construction are generally short-term, temporary, and reversible, which means they can be reduced or eliminated by known mitigation measures. The project is not likely to cause significant environmental impacts. The environmental risk will involve: (i) occupational health and safety (OHS) risks during construction and maintenance of the system; the occurrence of OHS incidents has low probability (possible but not likely) and could be minimized through abiding by tailored OHS plans for each activity; (ii) risks associated with handling hazardous wastes; and (iii) risks of handling hazardous substances and wastes in case of having a storage battery within the system. The probability of contaminating disposal sites (as per points ii and iii above) is not high. However, the implementation of the project may imply noise, dust, and waste handling during construction, possible risks to animals and plants, handling of wasted old electricity meters, handling of waste batteries and solar panels and heat and sunlight reflection from solar panels. There might be other safety issues that should be determined by screening the project site. Table 8.1 proposes a summary of negative impacts and mitigation measures.

Table 8.1 Summary of negative impacts and mitigation measures.

Terrestrial habitat alteration	Mitigation measures
Environmental	
Construction	• Installation of solar panels above existing vegetation to avoid land clearing • Revegetation of disturbed areas with native plant species • Management of construction site activities as described in relevant sections of the general EHS guidelines
Maintenance	• Implementation of an integrated vegetation management approach (IVM). The selective removal of tall-growing tree species and the encouragement of low-growing grasses and shrubs is a common approach to vegetation management in transmission line and pipe line rights of way. Alternative vegetation management techniques should be selected based on environmental and site considerations, including potential impacts to nontarget, endangered, and threatened species

Table 8.1 Summary of negative impacts and mitigation measures—cont'd

Terrestrial habitat alteration	**Mitigation measures**
	• Removal of invasive plant species, whenever possible, cultivating native plant species • Observing manufacturer machinery and equipment guidelines, procedures with regard to noise, and oil spill prevention and emergency response
Avian collisions and electrocution	• Maintaining 1.5 m (60 in.) spacing between energized components and grounded hardware or, where spacing is not feasible, covering energized parts and hardware • Retrofitting existing transmission or distribution systems by installing elevated perches and insulating jumper loops, placing obstructive perch deterrents (e.g., insulated Vs), changing the location of conductors, and/or using raptor hoods • Installing visibility enhancement objects such as marker balls, bird deterrents, or diverters
Hazardous materials Insulating oils and fuels	Mitigation measures • Disposal of hazardous materials in accordance with the waste management plan for the different wastes that will be generated within their geographical areas
Occupational health and safety	
Live power lines	• Only allow trained and certified workers to install, maintain, or repair electrical equipment • Deactivating and properly grounding live power distribution lines before work is performed on, or in close proximity to, the lines • Ensuring that live-wire work is conducted by trained workers with strict adherence to specific safety and insulation standards. Qualified or trained employees working on transmission or distribution systems

Continued

Table 8.1 Summary of negative impacts and mitigation measures—cont'd

Terrestrial habitat alteration	Mitigation measures
	should be able to achieve the following: - Distinguish live parts from other parts of the electrical system - Determine the voltage of live parts - Understand the minimum approach distances outlined for specific live line voltages - Ensure proper use of special safety equipment and procedures when working near or on exposed energized parts of an electrical system • Workers should not approach an exposed energized or conductive part even if properly trained unless: - The worker is properly insulated from the energized part with gloves or other approved insulation, or - The energized part is properly insulated from the worker and any other conductive object, or - The worker is properly isolated and insulated from any other conductive object (live-line work) • Where maintenance and operation is required within minimum setback distances, specific training, safety measures, personal safety devices, and other precautions should be defined in a health and safety plan • Workers not directly associated with power transmission and distribution activities who are operating around power lines or power substations should adhere to local legislation, standards, and guidelines relating to minimum approach distances for excavations, tools, vehicles, and other activities

Table 8.1 Summary of negative impacts and mitigation measures—cont'd

Terrestrial habitat alteration	Mitigation measures
	• Minimum hot stick distances may only be reduced provided that the distance remaining is greater than the distance between the energized part and a grounded surface
Working at height	• Test structures for integrity prior to undertaking work • Implementation of a fall protection program that includes training in climbing techniques and use of fall protection measures; inspection, maintenance, and replacement of fall protection equipment; and rescue of fall-arrested workers, among others • Establishment of criteria for use of 100% fall protection (typically when working over 2m above the working surface, but sometimes extended to 7m, depending on the activity). The fall protection system should be appropriate for the tower structure and necessary movements, including ascent, descent, and moving from point to point • Installation of fixtures on tower components to facilitate the use of fall protection systems • Provision of an adequate work-positioning device system for workers. Connectors on positioning systems should be compatible with the tower components to which they are attached • Hoisting equipment should be properly rated and maintained and hoist operators properly trained • Safety belts should be of not less than 16mm (5/8in.) two-in-one nylon or material of equivalent strength. Rope safety belts should be replaced before signs of aging or fraying of fibers become evident

Continued

Table 8.1 Summary of negative impacts and mitigation measures—cont'd

Terrestrial habitat alteration	Mitigation measures
	• When operating power tools at height, workers should use a second (backup) safety strap • Signs and other obstructions should be removed from poles or structures prior to undertaking work • An approved tool bag should be used for raising or lowering tools or materials to workers on structures
Electric and magnetic fields	• Identify potential exposure levels in the workplace, including surveys of exposure levels in new projects and the use of personal monitors during working activities • Train workers in the identification of occupational EMF levels and hazards • Establish and identify safety zones to differentiate between work areas with expected elevated EMF levels compared to those acceptable for public exposure, limiting access to properly trained workers • Implement action plans to address potential or confirmed exposure levels that exceed reference occupational exposure levels developed by international organizations such as the International Commission on Nonionizing Radiation Protection (ICNIRP), and the Institute of Electrical and Electronics Engineers (IEEE). Personal exposure monitoring equipment should be set to warn of exposure levels that are below occupational exposure reference levels (e.g., 50%). Action plans to address occupational exposure may include limiting exposure time through work rotation, increasing the distance between the source and the worker when feasible, or the use of shielding materials

Table 8.1 Summary of negative impacts and mitigation measures—cont'd

Terrestrial habitat alteration	Mitigation measures
Community health and safety	• Use signs, barriers (e.g., locks on doors, use of gates, use of steel posts surrounding transmission towers, particularly in urban areas), and education/public outreach to prevent public contact with potentially dangerous equipment • Surround conducting objects (e.g., fences or other metallic structures) installed near power lines to prevent shock
Electromagnetic interference	• Create emission line rights of way and conductor bundles to ensure radio reception at the outside limits remains normal
Visual amenity	• Extensive public consultation as indicated in the SEP during the planning of power line and power line right-of-way locations • Accurate assessment of changes in property values due to power line proximity • Siting power lines, and designing substations, with due consideration to landscape views and important environmental and community features • Burying transmission or distribution lines when power must be transported through dense residential or commercial areas
Noise	• Project planning stages to locate rights of way away from human receptors, to the extent possible. Use of noise barriers or noise canceling acoustic devices should be considered necessary
Social impacts and risks	
Labor conditions and community health and safety	• GBV, HIV/AIDS, child protection training/awareness campaign for contractors, subcontractors, and communities (and HIV/health) • Provisions for handling of GBV in the GRM

Continued

Table 8.1 Summary of negative impacts and mitigation measures—cont'd

Terrestrial habitat alteration	Mitigation measures
	• Update and implement the stakeholder engagement plan (SEP) • Communication through contractor's environmental and social specialist when stringing activities will take place to ensure children are not playing in the work area • Project sites to be marked off with fencing and signage to prevent people from entering the dangerous sites
Land acquisition, involuntary displacement and restrictions on land use	• Update the resettlement framework (RF) to resettlement action plan(s) (RAP) and implement it • Development of subsequent and/or livelihood restoration plans (LRP)

For such a project, the environmental and social risk is classified as follows: substantial, moderate, and low.

In general, all essential environmental and social instruments should be prepared. A list of these requirements that will be used as reference is shown below

- Environmental and social management framework.
- Waste management plan.
- Stakeholder engagement plan.
- An installation guide for PVPS systems.
- Labor management procedure.
- Resettlement framework.
- Grievance mechanism.

Generic environmental and social management plans (ESMPs) should be prepared to provide a logical framework within which identified negative environmental and social impacts can be mitigated and monitored. In addition, the ESMPs assign responsibilities of actions to various actors and provide a timeframe within which mitigation measures and monitoring can be done. All bidding documents should include E&S management measures and that all contracts will include subproject specific management plans to be implemented by contractors and subcontractors.

8.3 Environmental and social management procedure

The environmental and social management procedure of the proposed project starts with the identification of subprojects followed by social and environmental baseline screening of the subprojects. Based on the environmental and social screening, the generic ESMP will be updated to include relevant environmental and social issues. If a subproject is found to have no significant social safeguard issues,including loss of land, assets, or source of income, only a social safeguard report will be prepared summarizing the findings of the screening. However, if the screening identifies social safeguard issues, the RF will be updated to a resettlement action plan (RAP).

Environmental and social screening of each subproject and ESIA/RAP wherever required is to be subject to review and clearance by the bank. Whenever requested, the owner will provide copies of the filled out environmental and social screening forms for all subprojects to be implemented by the owner.

For an effective integration of environmental and social standards into the project implementation, the contractor will need to adopt this ESMP and prepare a comprehensive construction environment and social management plan (C-ESMP) that will provide the key reference point for compliance. The environmental supervision will also adopt the C-ESMP.

There should be an entity that is responsible for monitoring the overall implementation of the ESMP for all subprojects. In particular, the ESO will (i) monitor the implementation of mitigation measures and the environmental and social performance of contractors, and (ii) Monitor training of project staff, implementing partners, and contractors (list of persons, dates, and places).

The ESO will also prepare (i) quarterly reports summarizing monitoring results, to be included in the project's quarterly reports to the owner, (ii) reports that aggregate and analyze monitoring results ahead of regular implementation support missions with the owner, (iii) an annual evaluation of all environmental and social monitoring results, which will be submitted to the owner as part of overall project implementation reporting.

The engaged staff in general should be trained on awareness issues related to environmental management, policies, regulation, environmental assessment, and monitoring activities that are related to energy projects as well as recording and resolving any grievance matters related to the implementation of the project.

The cost estimated for ESMP implementation is related to hiring and training the ESO, beneficiaries and stakeholder training, environmental orientation seminars for contractors, and remuneration of an external local consultant to conduct an external monitoring of the resettlement process to support the owner in the implementation of the ESMP. Other indicative costs may be related to the implementation of stakeholder engagement and anticipated resettlements.

8.4 Public consultations and stakeholder engagement plan

The owner should prepare a stakeholder engagement plan proportional to the nature and scale of the project impacts and risks, the plan will need to be updated from time to time. The main stakeholders have been identified in the SEP and the need for their engagement throughout the project cycle has been outlined. The SEP details the enhanced requirement to engage with the project affected during the preparation and implementation of ESIA/ESMP and RAP. The project-specific grievance mechanism has also been detailed; see Section 8.19.6.

Several consultations should be undertaken for the preparation of the ESF documents, including this ESMP and the SEP. The stakeholder engagement activities should be conducted in different regions and locations with a variety of stakeholders. Moreover, the communities that are affected by the project should be informed during the meetings that a GRM system will be available prior to the implementation phase to file complaints and concerns related to the project activities. A summary of the different stakeholder engagement activities that have taken place to date and a detailed description of the stakeholder engagement activities for each component are available in the SEP document.

On the other hand, the resettlement Framework (RF) should be prepared by the owner to identify the project affected persons (PAPs), the types of impacts, and strategies for the compensation and/or restoration of potential losses for individuals and businesses. It is prepared in a separate document and it will be used as a basis for the preparation of the specific resettlement action plan, RAP, once the designs are completed and exact locations are identified. The RF sets out the policies, principles, institutional arrangements, schedules, and indicative budgets that will take care of anticipated resettlements for various project components.

In any ESMP, the term "associated Facilities" means facilities or activities that are not funded as part of the project and are: (a) directly and significantly related to the project; (b) carried out, or planned to be carried out, contemporaneously with the project; and (c) necessary for the project to be viable and would not have been constructed, expanded, or conducted if the project did not exist. If any associated facility will be identified during the progressive definition of the program components, the same will be subjected to screening in order to determine the gaps to be addressed under the environmental and social instruments.

The screening process will include site visits and interviews with operators to discuss the monitoring activities performed by the operation staff and their compliance to the monitoring program stated in the project ESMP. The screening process will also involve direct consultation with the PAP(s) and officials from district administration onsite to verify the affected assets and discuss their socioeconomic situation. Before the process begins, the PAP(s) should be advised in writing and verbally of their rights and will be consulted throughout the resettlement process. This will include sharing a copy of the grievance redress procedure and the entitlement matrix.

8.5 Applicability of projects

After having a full description of the project nature, type, activity, and implementation and operation processes, there should an assessment of these activities considering environmental and other related laws in selected locations.

8.5.1 Assessment and management of environmental and social risks and impacts

The analysis and assessment of the environmental and social risks and risk classification for each subproject will be carried out. Quantitative/qualitative analysis is carried out as much as possible to identify all project impacts, including direct and indirect, short-term and long-term, reversible and irreversible, and cumulative for the construction and operation phases of the project. A set of mitigation, monitoring, and institutional measures to be taken during the implementation and operation of a project to eliminate adverse environmental and social risks and impacts, offset them, or reduce them to acceptable levels, is prepared. Hence, a generic environmental and social management plan (ESMP) is prepared for each subproject, which includes the measures and actions needed to implement these measures.

The owner should (a) identify the set of responses to potentially adverse impacts; (b) determine requirements for ensuring that those responses are made effectively and in a timely manner; and (c) describe the means for meeting those requirements.

8.5.1.1 Labor and working conditions

Each project subcomponent is expected to employ a different number of labor workers during peak construction and operation. The labor workers will be contracted through the project contractor. A labor management procedure (LMP) is prepared to set out the project's approach to meeting national requirements as well as the objectives of the ESF concerning labor and working conditions. It is more practical to hire laborers from the target area.

8.5.1.2 Human resources policies and procedures

Appropriate human resources policies and procedures that are in line with the local labor law should be adapted by the OWNER. For this purpose, a working management procedure should be prepared.

A training plan will also be put in place for employees and contractors. Induction training on the HR policy and procedures and basic safety awareness training will be provided to all newly hired workers. Other types of technical skills training will be identified for staff on an as-needed basis. This includes training of technicians and laborers on how to install, operate, and maintain a PV solar plant and electrical transmission lines.

8.5.1.3 Working conditions and terms of employment

The project's HR policies and procedures will specify the terms of employment (wages and benefits, hours of work, overtime arrangements and overtime compensation, annual and sick leave, vacation and holiday, health insurance, and end-of-service benefits) and will also include provisions on restrictions to child labor and the prevention of forced labor as well as commitment to nondiscrimination and equal opportunities for employees and contractors; it will be shared with all new hires. Nondiscrimination and equal opportunity will be adopted for all workers.

8.5.1.4 Grievance mechanism

A workers' grievance mechanism should be developed and made available to all workers, including contractors and subcontractors. The grievance mechanism will, among other things, clearly define the response timeframes to grievances and incorporate a grievance log as part of the grievance redress mechanism process.

8.5.1.5 Occupational health and safety

Key occupational health and safety (OHS) risks for all project subcomponents include slips and falls, potential hazards from onsite moving machinery, heavy load lifting, exposure to electric shocks and burns, exposure to high-voltage lines, and safety issues related to PV module assembly. In addition to weather conditions such as a relatively hot project location, construction workers might be at risk of dehydration, heat exhaustion, and heat stroke if not properly hydrated.

Prior to the start of construction activities, the owner will ensure that occupational health and safety procedures exist and will cover the following issues: hazard identification and assessment; construction site safety (barricades, safety nets, access control, clear demarcation of areas, and provision of safety information to visitors); specific procedures for hazardous work; workers' safety and training plan; personal protective equipment needs; site supervision and audit procedures; and incident intervention measures and reporting. The procedure will be designed to be specific to different project components, including electrical transmission line connections, PV installation, and battery recycling factories. OHS procedures will be revised and updated for operations where the risk is reduced.

8.5.2 Resource efficiency and pollution prevention and management

8.5.2.1 Resource efficiency

Resource consumption on all project components is expected to be minimal, with the main resource utilized during construction being water for dust suppression, concrete production, and domestic usage. During operations, the main water use will be cleaning the PV modules and domestic usage.

8.5.2.2 Water consumption

Considering the aforementioned described components that include photovoltaic system installation including rooftop systems and photovoltaic system plants, the estimated water consumption for the estimated photovoltaic system capacity during the construction phase varies from component to component; for the PV panel construction component, it is not expected to be more than 20 m3/day. This includes water for drinking, site activities, and civil activities (concrete production, equipment cleaning, and dust suppression). During operations, water consumption is estimated to be $100\,m^3$/year for modules cleaning and $100\,m^3$/year for general sanitary use for the project. Wet cleaning technology will be used to ensure that dust and other particles accumulated on the panels do not compromise the efficiency of the PV facility. The cleaning cycle is estimated to be 12 cleanings every year. A water management plan will be developed for the project that will include provision of the required water quantity and reporting of water use and sanitation.

8.5.2.3 Greenhouse gases

Greenhouse gas emissions from the project during construction are expected to be predominantly associated with the use of fuels such as in generators, transport, onsite equipment, and machinery, although the emissions have not been calculated. For PV construction components, this is expected to be low and significantly less than 800 tons of CO_2 per year. These estimations are done considering the estimated size of the photovoltaic system installation based on project components that have been described before.

8.5.2.4 Wastes

Solid waste generated during construction mainly consists of municipal and construction wastes that will be collected by a local joint service council to be disposed of in an authorized landfill. The overall volumes of both solid and hazardous waste generated by the project during construction are expected to be low. It is anticipated that solid waste will comprise paper, wood, plastic, scrap metals, and glass. Hazardous waste will likely be comprised of fuel, oils, lubricants, hydraulic/insulating fluids, batteries, tires, metal drums, and empty chemical containers. Scrap metals, plastic, batteries, metal drums, old meters, and glass waste will be rewarded to small factories that recycle them. A waste management plan should be prepared for stakeholders and implemented for the safe management of this waste. Wood waste will be given to factories that reshape it and make it usable for fireplaces for residential purposes. A limited number of waste PV modules is expected to require disposal during the construction phase. During operations, waste generated will be largely limited to domestic waste and waste generated from maintenance. These waste streams will be segregated as per the waste management plan to be developed for construction. When the plant is decommissioned, the priority option of disposal of the PV panels will be according to EQA hazardous waste management regulations that meet Basel convention requirements.

8.5.2.5 Wastewater treatment

During the construction phase, wastewater from sanitary facilities will be stored in suitable septic tanks and transported offsite. The owner and the contractor will have the overall responsibility for management and assignment of proper firms for the management, collection, and disposal contracts for sewage and other wastewater from sites.

8.5.2.6 Pollution prevention

During project construction, power needs will be met via 100 k-volt-ampere (KVA) diesel generators. These will locally impact air quality and require fuel management and containment. These impacts, however, will be short in duration (maximum 6 months) and vary from component to component. During operations, electricity will be back-fed from the grid. Plans and procedures that manage the pollution-related aspects of the project's component will be in line with the requirements of relevant national regulations. Aspects should cover air quality/dust, spills, and occupational noise, among others.

8.5.3 Community health and safety

The safety of communities will be examined. The prepared ESMPs aim to minimize the risks during the construction phases. Wherever a construction activity requires route detours, safety measures for nearby communities and road users from traffic and construction activities will be put in place in accordance with acceptable norms as per the international EHS guidelines and enforced. It is essentaial that communities are not exposed to hazardous materials during the construction phase. There is a need to ensure that the safeguarding of personnel and property is carried out in a manner that avoids or minimizes risks to the project-affected communities. The design will also consider the health and safety of the communities close to the line after project implementation.

8.5.4 Land acquisition, restrictions on land use and involuntary resettlement

Some of the project's components may require land acquisition and restrictions on land use, which can have adverse impacts on communities and people. Project-related land acquisition or restrictions on land use may cause physical displacement (relocation, loss of residential land or loss of shelter), economic displacement (loss of land, assets, or access to assets, leading to loss of income sources or other means of livelihood) or both. Therefore, a resettlement framework (RF) is prepared.

Once the design is complete and impacts on the known site are identified, specific resettlement action plans, RAPs, if the number of affected is less than 200 per each subproject, will be prepared by the ESO based on the RF. Specific measures that consider their situation to ensure they will not be negatively impacted will be carried out.

8.5.5 Biodiversity conservation and sustainable management of living natural resources

An investigation of the relevance of this issue to the project activities indicates little relevance. Still, a site-specific environmental impact assessment will be carried out to assess the impacts on natural resources once exact locations of project components are available.

8.5.6 Indigenous peoples/sub-Saharan African historically underserved traditional local communities

A provision of this standard does not apply to the proposed project's components because there are no indigenous communities in the area. Vulnerable communities such as refugees and Bedouins, if any, that will be affected are covered. The owner should ensure that the development process fosters full respect for human rights, dignity, aspirations, identity, culture, and natural resource-based livelihoods and the tradition of local communities.

8.5.7 Cultural heritage

During the implementation of the project, cultural heritage components should be expected. Moreover, there should be no registered archeological sites within or in close proximity to the proposed project location. This will be confirmed by the site-specific safeguard instruments and chance-fined procedures will apply to all construction work that will comprise evacuation, demolition, or land movement.

8.5.8 Financial intermediaries

There are no actions under the project related to financial intermediaries, therefore there are no mitigation measures to be undertaken.

8.5.9 Stakeholder engagement and information disclosure

Project stakeholders should identified as direct and indirect stakeholders and beneficiaries.

8.5.10 Information disclosure

The owner should disclose project information and all key documentation, including this environmental and social management plan, to allow stakeholders to understand the risks and impacts of the project as well as potential opportunities. The information should be disclosed in relevant local languages and in a manner that is accessible and culturally appropriate, taking into account any specific needs of groups that may be differentially or disproportionately affected by the project or groups with specific information needs (such as disability, literacy, gender, mobility, differences in language or accessibility).

The disclosure should include information on: (i) the stakeholder engagement process, highlighting the ways in which stakeholders can participate; (ii) the time and venue of any proposed public consultation meetings, and the process by which meetings will be notified, summarized, and reported; and (iii) the process and means by which grievances can be raised and will be addressed.

8.6 Environmental and social risk and impacts

In general, the guidelines include parameters for environmental assessment, public consultations, and measures to enhance project benefits to communities and women. Together, these guidelines provide the methods to identify the environmental and social impacts associated with the implementation of subprojects and include measures to mitigate such problems as well as enhance environmental and social performance. The environmental and social assessment of the subprojects is to:

- Define the specific environmental and social instruments;
- Identify potential environmental and social impacts;
- Determine appropriate environmental risk categories, according to Bank's Environmental and Social standards;
- Review and approve subprojects;
- Identify any mitigation and monitoring indicator measures;

- A stakeholder engagement framework, a stakeholder engagement plan for specific sites, labor management procedures, ESCP, and a resettlement framework are prepared.

As for a photovoltaic water pumping system project, it is not likely to cause significant environmental impacts. The environmental risk is rated as substantial as some of the project activities involve: (i) occupational health and safety risks during construction and maintenance of transmission and distribution lines. The occurrence of OHS incidents has low probability (possible but not likely) and could be minimized through abiding by tailored OHS plans for each activity; (ii) the risks associated with handling hazardous wastes, such as waste oil from transformers during maintenance; and (iii) the risks of handling hazardous substances and wastes during upgrading the battery. The probability of contaminating disposal sites (as per points ii and iii above) is also low (possible but not likely) and could be minimized by employing adequate waste management procedures. Other environmental risks are mainly moderate and low including noise, dust, and waste handling during construction exposure to electromagnetic fields (EMF), possible risks to birds, handling of wasted old electricity meters, and handling of waste batteries, solar panels, and heat and sunlight reflection from solar panels. It is worth noting that as an MPA, this initial rating would help improve the overall environmental and social risk management practices with the owner, and, with close supervision by the team, the capacity of the clients for handling the environmental and social risks will be boosted.

8.6.1 Guidelines on environmental and social risk classifications

The ESS applicability to the subproject will result in recommending the appropriate needed environmental and social management documents/ instruments such as ESIA, ESMP, SEP, and RAP in compliance with international standards. If the risk rating of a subproject increases to a higher risk rating, the owner should apply relevant requirements. The measures and actions agreed should be included in the ESCP and monitored.

Below, more details of environmental classifications of which the subprojects will be classified accordingly are presented.

(a) High risk classification: A proposed project is classified as high risk if it is likely to have significantly adverse environmental impacts. These impacts may affect an area broader than the sites or facilities subject to physical works. A full EIA is required. The EIA examines the project's potential negative and positive environmental impacts; compares them with those of feasible alternatives, including a no-action, that is, no-project alternative; and also incorporates public consultations as per the national EIA regulation requirements. The EIA will recommend needed measures to prevent, minimize, mitigate, or compensate for adverse impacts and help improve environmental performance.

(b) Substantial risk classification: A proposed project is classified as substantial risk if its potential adverse environmental impacts on human populations and the environment are less adverse than those of the high risk category. These impacts are site-specific; few if any of them are irreversible; and in most cases mitigation measures can be designed and implemented more readily than for high risk category projects.

(c) Moderate or low risk classification: A proposed project is classified as moderate or low risk if it is likely to have minimal or no adverse environmental impacts. For projects with moderate impacts and a limited scale/footprint, an ESMP will be required. For projects of low risk, simple mitigation measures in place such as a checklist or no further environmental action are required for this classification.

8.7 Assessment of environmental and social impacts of PVPS

Based on the social and environmental baseline, the following likely environmental and social impacts and their risk classifications can be drawn:

Noise emissions: The construction and installation phase could generate increased noise and disturbance to the surrounding properties and neighbors. The allowed dB level of the used equipment on operations will be according to the vicinity to the residential area.

Air emissions: GHG emissions, caused by fossil fuel sources used in production, manufacturing, waste disposal, and recycling, are embodied in renewable technologies. GHG reduction will be evaluated once implementation is commissioned.

Electromagnetic fields (EMF): During the operation phase, the public may be exposed to electric and electromagnetic fields.

Hazardous chemicals: Hazardous chemicals such as boron and phosphorus are often used in the production of solar modules. However, these do not pose a danger during the operational phase and only become dangerous during disposal; hence, a waste management plan is required.

Impacts on biodiversity: Generally, the rehabilitation of the connection points does not have adverse impacts on biodiversity. If the construction of a new connection point requires tree trimming or cutting, and it is determined that the tree is not a protected species, it might warrant cutting the tree. A mitigation measure for such an issue is to plant replacement trees in an agreed-upon site with the tree owner. When trees are removed, they must be planted in the same type of soil in order to ensure their continuous growth.

Heat or light reflection: Neighboring properties could be affected by the reflection of sunlight from the panels, especially if angled toward neighboring windows, doors, or balconies. The contractor should follow proper standards for installation. If this reflection is sustained for a prolonged period, it may become a source of grievance.

Cultural heritage: A proposed intervention should not affect any buildings or places that are deemed to have significant cultural values. However, in the course of project implementation, a chance find may occur whereby historical and cultural property is inadvertently found. Chance find procedure clauses for avoiding potential impacts will be inserted into the civil works contracts to ensure that the necessary measures are in place during the construction phase of the subproject.

Employment: The installation of the domestic solar systems will help address high youth unemployment in the area by providing temporary employment for technical and unskilled labor.

Economic impact and livelihoods: Rooftop solar installations will help to stabilize the electricity supply to beneficiary households while making use of an abundant raw material–sunlight. This could result in cost savings, an improved standard of living, and increased household income. It could also contribute to increasing business operations and thus revitalizing businesses that depend on electricity to function.

Social conflicts: The likelihood for social conflicts is minimal and manageable. Potential conflicts could arise from fairness and equity in decision-making and the use of nonlocal manpower during project implementation (installation). To the extent possible, local manpower should be employed to deploy the installations.

Occupational health and safety: This subproject is associated with considerable risks to occupational health and safety during construction and maintenance. Each contractor should present an OSH plan including risk assessment and mitigation measures.

Waste disposal: All waste generated from the project before, during, and after installation must be disposed of at a designated disposal site in coordination with the relevant authority. Any hazardous materials must be disposed of under the guidance of the EQA.

This component requires a resettlement framework. In a case involving land, its possible that a site-specific resettlement action plan (RAP) will be needed. Site-specific RAPs, if needed, will be disclosed before the commencement of work with the possible exception of livelihood initiatives. The SEP will be updated to include site-specific information and consultation. An LMP is required for this subproject. The contractor will include particular provisions on occupational health and safety measures, child labor, and work conditions including forced labor, following occupational health and safety guidelines. It should be dealt with in accordance with national laws and ESSs.

8.8 Environmental and social instruments

Because most subproject locations are not known yet, this ESMF is prepared. The ESMP should identify the applicability of different ESSs to the project activities as clarified above.

Environmental and social impact assessment (ESIA): An ESIA should be prepared for photovoltaic water pumping projects. The ESIA assesses the different risks and impacts related to this subproject (rated as of substantial risk due to the OHS risks and possible handling of hazardous substances during construction and operation). The ESIA presents the baseline conditions along the identified route, and concludes that there are no environmentally sensitive areas, especially in terms of biodiversity and avian fauna.

Labor management procedure (LMP): The LMP is prepared to set out the project's approach to meeting national requirements as well as the objectives of the bank's ESF, specifically objectives of the environmental and social standard.

Stakeholder engagement framework (SEF): An SEF is prepared. For subprojects where their locations are unknown, a stakeholder engagement framework (SEF) has been prepared, which will be the basis for future package SEPs once the packages are identified.

Resettlement framework (RF): The RF should be prepared in case a project is in a location where Bedouin communities are located or if the project affects any living communities nearby.

8.9 Environmental and social management procedure for PVPS

All project components or subprojects to be implemented under the proposed project will be subject to an environmental/social screening in order to prevent execution of projects with significant negative environmental impacts. The purpose of environmental/social screening is to get a preliminary idea about the degree and extent of potential environmental impacts of a particular subproject, which would subsequently be used to assess the need for further environmental/social assessment. The subprojects will be identified by the owner after selection of a subproject. The ESO will conduct environmental/social screening of the subproject as an integral part of planning and implementation.

The environmental and social screening process will determine the nature of the environmental and social assessment that should be subsequently undertaken. The environmental/social screening will provide a rapid assessment of the project characteristics, its beneficiaries, the socioeconomic dimensions of the area, and its potential environmental/social impacts and risks. The results of the environmental/social screening will determine whether a subproject requires further environmental and social assessment, including an ESIA and a RAP.

The environmental/social screening would involve: (i) reconnaissance of the subproject areas/routes and their surroundings; (ii) identification of the major subproject activities; and (iii) preliminary assessment of the impacts of these activities on the ecological, physicochemical, and socioeconomic environment of the subproject surrounding areas.

The ESO would carry out the "environmental/social screening" of subprojects with a preliminary idea about the nature of the subproject location and subproject activities by filling in the for solar system. Based on the evaluation results, the ESO should follow the ESMP procedure so as to avoid environmental and social risks of the project.

8.10 Environmental and social management plan for PVPS

The ESMP consists of the set of mitigation, monitoring, and institutional measures to be taken during the implementation and operation of a project to eliminate adverse environmental and social risks and impacts, offset them, or reduce them to acceptable levels. The EMP also includes the measures and actions needed to implement them. The owner will:

- Identify the set of responses to potentially adverse impacts;
- Determine the requirements for ensuring that those responses are made effectively and in a timely manner; and
- Describe the means for meeting those requirements.

8.10.1 Mitigation and enhancement measures

Depending on the project, an ESMP is prepared and incorporated directly here. This ESMP includes mitigation measures as well as responsibilities for planning, implementation, supervision, and monitoring. Hence, the following environmental management plans describe how an action might impact the natural environment in which it occurs and set out clear commitments from the person taking the action on how those impacts will be avoided, minimized, and managed so that they are environmentally acceptable. The ESMP is targeted to exclude the deterioration of the environment during the implementation of planned activities.

The owner should mitigate the environmental and social impacts associated with construction activities by: (i) including environmental and social clauses in all supply and installation/construction contracts and (ii) ensuring that contractor personnel are familiar with these clauses.

Mitigation measures that will effectively address potential risks and impacts as well as implementation and monitoring responsibilities for mitigation measures are identified. Table 8.2 provide generic ESMP matrices.

8.10.2 Monitoring plan

Based on the above EMP and the mitigation measures set, the following monitoring plan is developed to help, track, and assess the interventions set in the mitigations of each environmental parameter throughout the life cycle of the project. In this plan, all indicators to be monitored and the institutional responsivities are presented, in addition to a column showing the implementation route/plan needed (Table 8.3).

8.11 Occupational health and safety guidelines

In general, it is requested to follow the international environmental health and safety and guidelines for electric power transmission and distribution as they are stricter. A summary of essential safety mitigation measures that should be applied during the implementation of the project includes:

Table 8.2 Generic ESMP matrix.

Project activity	Potential environmental/ social issue	Management/mitigation measures	Responsibility: Implementation	Responsibility: Supervision and monitoring
Construction stage				
Right-of-way maintenance	Biodiversity	- Replant the trees in the same type of soil. - Avoid installation sites that require cutting or substantially pruning a protected tree, an old tree, or a known bird-nesting tree	Contractor	Owner
	Risk of cut vegetation burning	- Avoid disposing of any cut materials by burning	Contractor	Owner
	Risk of damaging habitats and species outside the RoW by the workers	- Prevent habitats and species damage outside the RoW by the workers	Contractor	Owner
	Impacts on private lands	- Prevent any close alignments of the routes that may cross private lands	Contractor	Owner
Excavation and civil work	Noise impact	- Implement the installation activities only during the daytime - Inform the neighborhood of the activities before any installation - Use noise canceling devices	Contractor	Owner
	Dust	- Obviate large machines - Inform the neighborhood of the activities	Contractor	Owner

	Excavation for towers and outdoor rooms could damage water pipes in the village	- Avoid any damage to the existing infrastructure by consulting the local community	Contractor	Owner
	Waste disposal	- Dispose packaging and construction waste at approved waste management sites using registered transport services - Use untreated waste as domestic waste	Contractor	Owner
Installation of electrical components	Exposure to electrical and magnetic fields	- Put warning signs in the areas where the electrical and magnetic fields are high - Train the workers of the occupational electric and magnetic levels and other hazards	Contractor, safety inspector	Owner
	Chemicals and hazardous materials used in the switchgear	- Handle products carefully to avoid accidental breakage or spillage - Adopt safety procedures and personnel safety equipment to handle any chemicals used such as SF6	Contractor, safety inspector	Owner
	Occupational health and safety (working at height)	- Test structures for integrity prior to undertaking work - Implementation of a fall protection program that includes training in climbing techniques and use of fall protection measures; inspection, maintenance, and replacement of fall protection equipment; and rescue of fall-arrested workers, among others - Establishment of criteria for use of 100% fall protection (typically when working over 2 m above the working surface, but sometimes extended to 7 m, depending on the activity). The fall protection system should be appropriate for the tower structure and necessary movements, including ascent, descent, and moving from point to point	Contractor, safety inspector	Owner

Continued

Table 8.2 Generic ESMP matrix—cont'd

Project activity	Potential environmental/social issue	Management/mitigation measures	Responsibility: Implementation	Responsibility: Supervision and monitoring
		- Installation of fixtures on tower components to facilitate the use of fall protection systems - Provision of an adequate work-positioning device system for workers. Connectors on positioning systems should be compatible with the tower components to which they are attached - Hoisting equipment should be properly rated and maintained and hoist operators properly trained - Safety belts should be of not less than 16 mm (5/8 in.) two-in-one nylon or material of equivalent strength. Rope safety belts should be replaced before signs of aging or fraying of fibers become evident - When operating power tools at height, workers should use a second (backup) safety strap - Signs and other obstructions should be removed from poles or structures prior to undertaking work - An approved tool bag should be used for raising or lowering tools or materials to workers on structures		
	Occupational health and safety (live power lines)	• Only allow trained and certified workers to install, maintain, or repair electrical equipment	Contractor, safety inspector	Owner

		• Deactivating and properly grounding live power distribution lines before work is performed on, or in close proximity, to the lines • Ensuring that live-wire work is conducted by trained workers with strict adherence to specific safety and insulation standards. Qualified or trained employees working on transmission or distribution systems should be able to achieve the following: – Distinguish live parts from other parts of the electrical system – Determine the voltage of live parts – Understand the minimum approach distances outlined for specific live line voltages – Ensure proper use of special safety equipment and procedures when working near or on exposed energized parts of an electrical system • Workers should not approach an exposed energized or conductive part even if properly trained unless: – The worker is properly insulated from the energized part with gloves or other approved insulation; or, – The energized part is properly insulated from the worker and any other conductive object; or, – The worker is properly isolated and insulated from any other conductive object (live-line work). • Where maintenance and operation are required within minimum setback distances, specific training, safety measures, personal safety devices, and other precautions should be defined in a health and safety plan.		

Continued

Table 8.2 Generic ESMP matrix—cont'd

Project activity	Potential environmental/social issue	Management/mitigation measures	Responsibility: Implementation	Responsibility: Supervision and monitoring
		• Workers not directly associated with power transmission and distribution activities who are operating around power lines or power substations should adhere to local legislation, standards, and guidelines relating to minimum approach distances for excavations, tools, vehicles, and other activities; - Minimum hot stick distances may only be reduced provided that the distance remaining is greater than the distance between the energized part and a grounded surface.		
Social and cultural impacts	Employment of local people	- Employ local residents in the work	Contractor	Owner
	Risk of worker and local community harm from the accidents on site.	- Inform people of the activities to exclude them from all construction locations. - Keep any reports or records of the accidents. - Provide health and safety training for all personnel - Follow documented procedures for all site activities	Contractor, safety inspector	Owner
	Possible small pieces of land acquisition for some of the new towers	- Complete all necessary land acquisition in accordance with RAP and entitlement framework prior to the commencement of any construction works. - Ensure that the affected persons are: - Informed about their options and rights;		Owner

		- Consulted on, offered choices among, and provided with alternatives; - Provided prompt and effective compensation at full replacement cost for losses of assets attributable directly to the project; - Provided with development assistance in addition to compensation measures.			
Operation and maintenance					
Operation and maintenance as well as provision	Electrocution risk	- Use signs as well as barriers such as locks on doors and education/public outreach to prevent public contact with potentially dangerous equipment. - Surround conducting objects (e.g., fences or other metallic structures) installed near power lines to prevent shock.			Owner
Potential environmental issue	Management/mitigation measures			Responsibility	
	Preconstruction	Construction	Operation	Planning and implementation	Supervision and monitoring
Air emissions/dust	Identify suppliers of ISO or best industry standard-compliant products Minimize cut and fill operations, the site clearing and grubbing operations should be limited to specific locations only	Dust generation due to vehicle movement on haul roads/access shall be controlled through regular water sprinkler system		Contractor	ESO at the owner

Continued

Table 8.2 Generic ESMP matrix—cont'd

Project activity	Potential environmental/ social issue	Management/mitigation measures		Responsibility: Implementation	Responsibility: Supervision and monitoring
Noise emissions	–	1. Daytime installation activities only 2. Inform neighbors of work schedule	Undertake maintenance activities only during daytime	Contractor	ESO at the owner
Chemicals	Identify suppliers (PV panels, inverters, and batteries) of ISO- or best industry standard-compliant products	Handle products carefully to avoid accidental breakage or spillage	If the roof is used for rainwater harvesting, check the panels frequently for any damage.	Contractor	ESO at the owner
Heat or Light Reflection	Avoid installation sites that would require panels to be placed in a manner that would reflect light into an immediate neighbor's window, balcony or door for more than 30 days a year	Install screens to prevent light from reaching an immediate neighbor's window, balcony, or door	Same as construction stage if a new building is constructed next to the site following installation.		ESO at the owner

Biodiversity	Avoid installation sites that require cutting or substantially pruning a protected tree, an old tree, or known bird-nesting tree. In some specific sites, trees are to be removed in order to create open un-shaded area for PV panel installation	If a tree needs to be pruned, only remove parts that are absolutely necessary. Do not remove a mature tree unless absolutely necessary	Same as construction stage		ESO at the owner
Cultural heritage	Avoid selecting installation sites that are culturally or religiously sensitive	–	–		ESO at the owner
Employment	Train local workers as much as possible	Use local labor for skilled and semi-skilled labor Provide temporary employment for technical and unskilled labor	Same as construction phase		ESO at the owner
Visual impact	Public consultation during the planning of solar PV design			Contractor	ESO at the owner

Continued

Table 8.2 Generic ESMP matrix—cont'd

Project activity	Potential environmental/ social issue	Management/mitigation measures		Responsibility: Implementation	Responsibility: Supervision and monitoring
Economic impacts and livelihoods	Ensure wide dissemination of information to all stakeholders Prepare SEP	–	Ensure project performance information results are widely shared		ESO at the owner
Social conflict	1. Ensure fair competition by creating a level playing field. 2. Ensure access to information and transparency in decisions 3. Undertake public consultation and information dissemination and finally establish and create awareness grievance redness mechanism	Create awareness on grievance redress procedure	Same as construction phase		ESO at the owner

Occupational health and safety	–	1. Respect all safety measures required for working on rooftops. Use safety nets in roofs facing roads to prevent debris accidentally falling to the street. 2. Place appropriate warnings on the road. 3. Ensure compliance with the LMP so that: - All workers are above 18 years - Regular review and inspection of working sites - The contractor shall make available all IDs of his working team - Raise awareness and promote children's rights	–	Contractor, safety inspector	ESO at the owner
Waste disposal	Identify suppliers (PV panels, inverters, and batteries) or ISO or best industry standard compliant products Prepare waste management plan	Dispose of packaging and construction waste at approved waste management sites using registered transport services Use untreated waste as domestic waste Prepare installation guide for PV grids	Provide a temporary storage facility to contain waste ahead of final disposal to EQA approved facility Contact with recycling or waste disposal facility capable of handling solar panel and battery waste		ESO at the owner

Table 8.3 Monitoring plan.

Project activity/ aspect	Parameter	Indicator	Institutional responsibility			Project phase	Monitoring cost
			Implementation route/plan	Monitoring responsibility	Frequency		
Impact on flora	Visual inspection	Bare soil/soil erosion	ESMP	Contractor project manager/ supervising engineer	Monthly	Construction and operation	Included in supervision scope and costs
Air emissions and quality of dust	TPS, SO_2, CO, H_2S, CO_2, dust fallout	Bad odor, use of PPE, health and safety plan in use Record of induction for workers, active dust suppression	ESMP	Contractor/ supervising engineer	Monthly	Construction and operation	Included in supervision scope and costs
Safeguarding community health and safety	Visual inspection incident and accident records	Induction training records Safety working procedure Maintenance of complaints log and resolution process Evidence of effective GM Photographs of	SEP project performance grievance mechanism	Contractor/ supervising engineer	Daily	Prior to andduring construction and operation	Included in supervision scope and costs

		appropriate fencing; and signage around site perimeter and where identified through risk assessment process.					
Safeguarding worker occupation health and safety	Health and safety records Visual inspection Active and passive monitoring	Audits of PPE use, maintenance of disciplinary records, etc. Records of inductions, trainings, and toolbox talks Good "housekeeping" onsite Worker grievance records and resolution	OHS management system	Contractor/ supervising engineer	Daily	Construction and operation	Included in supervision scope and costs
Storage of hazardous materials and chemicals	Spillages visual inspection	MSDS for all stored chemicals Functioning storage containers Chemical usage records	Waste management plan	Contractor/ supervising engineer	Monthly audit review	Construction	Included in supervision scope and costs

Continued

Table 8.3 Monitoring plan—cont'd

Project activity/aspect	Parameter	Indicator	Institutional responsibility			Project phase	Monitoring cost
			Implementation route/plan	Monitoring responsibility	Frequency		
Traffic concerns	Visual inspection	Record of accidents involving project vehicles Banks men shall be used to direct vehicle traffic around construction sites and hazards during working hours (health and safety plan) Plan approved by project manager barrier and signage	Traffic management plan	Contractor/supervising engineer	Daily	Construction and operation	Included in supervision scope and costs
Public awareness and community perceptions	Community consultations	Grievance management records and resolution process Evidence of occurrence-event report	SEP GM	Contractor/supervising engineer	Monthly	Construction and operation	Included in supervision scope and costs

Noise	dB(A)	Measure included in design and procurement plans Hearing protection and PPE in use Record of equipment maintenance	ESMP	Contractor/ supervising engineer	Monthly	Construction and operation	Included in supervision scope and costs
Soil erosion	Visual inspection	Bare soil pillars	ESMP	Contractor/ supervising engineer	Weekly	Construction and operation	Included in supervision scope and costs
Solid waste management	Domestic refuse, metallic scraps	Documented approvals for placement of wastes	Comprehensive waste management plan	Contractor/ supervising engineer	Daily	Construction and operation	Included in supervision scope and costs
Land Acquisition, displacement, and restrictions on land use	Consultations Site visits	Records of compensation completion and completion rate Progress on RAP/land restriction plan (LRP) implementation Compliance with RPF/RAP and national legislation	RF/RAP	The owner	Daily	Prior to and during construction	Included in supervision scope and costs

Continued

Table 8.3 Monitoring plan—cont'd

Project activity/ aspect	Parameter	Indicator	Institutional responsibility			Project phase	Monitoring cost
			Implementation route/plan	Monitoring responsibility	Frequency		
Cultural heritage	Visual inspection	Records of CFP activated	ESMP (chance finds procedures)	Contractor/ supervising engineer	Daily	Prior to and during construction	Included in supervision scope and costs
Supply Chain	Reporting	Bidding documents and contracts Supply chain performance on ESS2 compliance	ESMP bidding documents	Contractor/ supervising engineer	Weekly	Construction and operation	Included in supervision scope and costs

8.11.1 Workers occupational health and safety

The following guidelines must be followed:

- Staff training and regular equipment service and testing
- Only trained and certified workers are permitted to install, maintain, or repair electrical equipment;
- Testing structures for integrity prior to undertaking work;
- Workers not directly associated with power transmission activities who are operating around power lines should adhere to local legislation, standards, and guidelines relating to minimum approach distances for excavations, tools, vehicles, pruning, and other activities
- Use of signs, barriers, and education/public outreach to prevent public contact with potentially dangerous equipment;
- Ensure provision and proper use of personal protective equipment (e.g., safety harness, helmet, dust masks, etc.);
- Follow safe work procedures;
- Community policing should be encouraged to reduce vandalism of towers;
- Ensure there is no encroachment on the transmission line way leave.

8.11.2 Community occupational health and safety

The following guidelines must be followed:

- Adherence to OSHA 2007 Act and its subsidiary legislation to ensure that the health and safety of the immediate neighbors and the public is not threatened.
- The contractor should ensure that construction work is undertaken in a manner that is not likely pose risks to community health and safety.
- The contractor should use barricading tape to prevent the public from accessing excavated tower foundations and work sites during construction.
- The contractor should put in place adequate hazard communication to the public by using appropriate signage as prescribed by national law and international best practices.
- The contractor should conduct public awareness sessions on safety requirements within construction sites.
- Adequate security where necessary for the public and staff should be provided.
- Public awareness of the public health issues should be identified.
- Working sights should be condoned and controlled access must be ensured.

8.11.3 Fall from heights

Contractors should follow safe work procedures and enforce the proper use of necessary protective equipment.

- Test structures for integrity prior to undertaking work;
- Implementation of a fall protection program that includes training in climbing techniques and the use of fall protection measures; inspection, surface, but sometimes extended to 7 m, depending on the activity). The fall protection system should be appropriate for the tower structure and necessary movements, including ascent, descent, and moving from point to point;
- Installation of fixtures on tower components to facilitate the use of fall protection systems;
- Provision of an adequate work-positioning device system for workers. Connectors on positioning systems should be compatible with the tower components to which they are attached;
- Hoisting equipment should be properly rated and maintained and hoist operators properly trained;
- Safety belts should be of not less than 16 mm (5/8 in.) two-in-one nylon or material of equivalent strength. Rope safety belts should be replaced before signs of aging or fraying of fibers become evident;
- When operating power tools at height, workers should use a second (backup) safety strap;
- Signs and other obstructions should be removed from poles or structures prior to undertaking work;
- An approved tool bag should be used for raising or lowering tools or materials to workers on structures.

8.11.4 Live power lines

Contractors/workers should also follow safe work procedures and equipment regarding working with live power lines including:

- Only allow trained and certified workers to install, maintain, or repair electrical equipment;
- Deactivating and properly grounding live power distribution lines before work is performed on, or in close proximity to, the lines;
- Ensuring that live-wire work is conducted by trained workers with strict adherence to specific safety and insulation standards. Qualified or trained employees working on transmission or distribution systems should be able to achieve the following:

  - Distinguish live parts from other parts of the electrical system.
  - Determine the voltage of live parts.
  - Understand the minimum approach distances outlined for specific live line voltages.
  - Ensure proper use of special safety equipment and procedures when working near or on exposed energized parts of an electrical system.
- Workers should not approach an exposed energized or conductive part even if properly trained unless:
  - The worker is properly insulated from the energized part with gloves or other approved insulation, or
  - The energized part is properly insulated from the worker and any other conductive object, or
  - The worker is properly isolated and insulated from any other conductive object (live-line work).
- Where maintenance and operation are required within minimum setback distances, specific training, safety measures, personal safety devices, and other precautions should be defined in a health and safety plan;
- Workers not directly associated with power transmission and distribution activities who are operating around power lines or power substations should adhere to local legislation, standards, and guidelines relating to the minimum approach distances for excavations, tools, vehicles, and other activities;
- Minimum hot stick distances may only be reduced provided that the distance remaining is greater than the distance between the energized part and a grounded surface.

8.12 Special environmental clauses (SECs) for tender document

8.12.1 Environmental and social clauses for contractors

The owner should incorporate the following standardized environmental and social clauses in tender documentation and contract documents so that potential bidders are aware of the environmental and social performance requirements expected from them, are able to reflect that in their bids, and are required to implement the clauses for the duration of the contract. The owner should enforce compliance by contractors with these clauses.

The clauses cover four issues:

- Environment, health, and safety (EHS).
- Environmental and social monitoring by the contractor.

- Environmental and social liabilities.
- Grievance mechanism for workers.

8.12.2 Environment, health, and safety

There are clauses for contractors that address environmental, health, and safety concerns. The purpose of the environment, health, and safety (EHS) clauses for contractors is to define minimum standards of practice acceptable to the owner. The clauses will be included in the bidding documents and contracts.

8.12.3 Contractor environmental and social management plan

Prior to the start of implementation of the specific subcomponent, each contractor must prepare and submit a contractor environmental and social management plan (C-ESMP) to the owner for acceptance. The C-ESMP will provide a detailed explanation of how the contractor will comply with the EHS clauses for contractors and demonstrate that sufficient funds are budgeted for that purpose and sufficient capacity is in place to oversee, monitor, and report on C-ESMP performance. The C-ESMP must include specific mitigation measures based on the project's environmental and social management plan, the final design, the proposed work method statements, and the nature of the project site. The C-ESMP should include management plans that cover the following issues:

8.12.4 Gender-based violence

Contractors must address the risk of gender-based violence through:

1. Mandatory and repeated training and awareness raising for the workforce about refraining from unacceptable conduct toward local community members, specifically women;
2. Informing workers about national laws that make sexual harassment and gender-based violence a punishable offense that is prosecuted;
3. Introducing a worker code of conduct as part of the employment contract that includes sanctions for noncompliance (e.g., termination);
4. Adopting a policy to cooperate with law enforcement agencies in investigating complaints about gender-based violence.

8.12.5 Child labor

Contractors must follow LMP and should not employ workers below the age of 18 in all project components.

8.12.6 Labor influx

Where contractors and labor come from outside the local area, contractors will need to maintain labor relations with local communities through labor codes of conduct.

8.12.7 Road

In order to carry out the construction work, the owner may close or divert certain specified roads, either permanently or temporarily. The contractor should arrange for alternative routes for transportation and/or pedestrians.

After breaking up, closing, or otherwise interfering with any street or footpath to which the public has access, the contractor shall make such arrangements as may be reasonably necessary so as to cause as little interference with the traffic in that street or footpath during implementation of the construction work as shall be reasonably practicable. Wherever construction works interfere with existing public or private roads or other ways over which there is a public or private right of way for any traffic, the contractor shall construct diversion ways wherever possible.

8.12.8 Movement of trucks

The contractor moving solid waste materials shall take strict measures to minimize littering of roads by ensuring that vehicles are licensed and loaded in such a manner as to prevent falling off or spilling of construction materials and by sheeting the sides and tops of all vehicles carrying mud, sand, and other materials and debris. Construction materials should be brought from registered sources in the area and debris should be transferred to assigned places in landfills with documented confirmation.

8.12.9 Traffic safety measures

The contractor shall provide, erect, and maintain traffic signs, road markings, barriers, traffic control signals, and other measures that may be necessary to ensure traffic safety around construction sites.

The contractor shall not commence any work that affects public roads and highways until all traffic safety measures necessitated by the work are fully operational.

8.12.10 Access to project sites

The contractor shall take all reasonable precautions to prevent or reduce any disturbance or inconvenience to the owners, tenants, or occupiers of adjacent properties, and to the public generally. The contractor shall maintain any existing right of way across the whole or part of the construction site and public and private access to adjoining frontages in a safe condition and to a standard not less than that pertaining at the commencement of the contract. If required, the contractor shall provide acceptable alternative means of passage or access to the satisfaction of the people affected.

8.12.11 Noise and dust control

The contractor shall take all practicable measures to minimize nuisance from noise and dust caused by collection equipment. This includes:

- Respecting normal working hours in or close to residential areas.
- Maintaining equipment in a good working order to minimize extraneous noises from equipment movement as well as emissions or fumes from the equipment.
- Shutting down equipment when it is not directly in use.
- Using operational noise mufflers if needed.
- Providing spray water when required to minimize the impact of dust.
- Limiting the speed of equipment used for waste collection.

8.12.12 Protection of the existing installations

The contractor shall properly safeguard all buildings, structures, works, services, or installations from harm, disturbance, or deterioration during the concession period. The contractor shall take all necessary measures required for the support and protection of all buildings, structures, pipes, cables, sewers, and other apparati during the construction period, and to repair any damage that occurs in coordination with concerned authorities.

8.12.13 Protection of trees and other vegetation

The contractor shall avoid the loss of trees and damage to other vegetation wherever possible. Adverse effects on green cover within or in the vicinity of construction sites shall be minimized. The contractor will restore vegetative cover, where feasible.

8.12.14 Cultural resources

The contractor will train construction crews and supervisors to spot potential archeological finds. In the event of a potential find, the contractor will inform the owner, who will in turn liaise with the national museum or a local university for quick assessment and action.

8.12.15 Clean-up of sites upon completion of work

The contractor shall clean up all sites before starting and after completing the work to remove oil and waste properly with environmentally good practices and safe disposal following hygiene procedures.

8.12.16 Worker health and safety

To avoid work-related accidents and injuries, the contractor will:

- Provide occupational health and safety training to all employees involved in the work. Provide protective masks, helmets, overalls, safety shoes, and safety goggles, as appropriate.
- Provide workers in high noise areas with earplugs or earmuffs.
- Ensure the availability of first aid boxes.
- Provide employees with access to toilets and potable drinking water.
- Provide safety and occupational safety measures to workers
- Properly dispose of solid waste at designated permitted landfill sites allocated by the local authorities.
- Carry out all procedures to prevent leakage of generator oil into the site.
- Ensure that the head of the well is covered tightly.
- Provide secondary tanks for oil and grease to avoid spills.

8.12.17 Site construction safety and insurance

To further enforce the compliance of environmental management, contractors are responsible and liable for the safety of site equipment, labor, and daily workers attending to the construction site as well as the safety of citizens for each project site, as mandatory measures.

8.12.18 Environmental and social monitoring by contractors

The owner should require that contractors monitor, keep records, and report on the following environmental and social issues for the project. The application of this requirement will be proportionate to the activities and to the size of the contract in an acceptable manner, including:

Safety: hours of work, recordable incidents, and corresponding root cause analysis (lost time incidents, medical treatment cases), first aid cases, high potential near misses, and remedial and preventive activities required (for example, revised job safety analysis, new or different equipment, skills training, and so forth).

Environmental incidents and near misses: environmental incidents and high potential near misses and how they have been addressed, what is outstanding, and lessons learned.

Major works: those undertaken and completed, progress against project schedule, and key work fronts (work areas).

E&S requirements: noncompliance incidents with permits and national law (legal noncompliance), project commitments, or other E&S requirements.

E&S inspections and audits: by contractor, engineer, or others, including authorities, to include the date, inspector or auditor name, sites visited, records reviewed, major findings, and actions taken.

Workers: the number of workers, indication of origin (expatriate, local, nonlocal nationals), gender, skill level (unskilled, skilled, supervisory, professional, management), and age with evidence that no child labor will be involved in all project components, including for rooftop PV systems, small-scale battery recycling, and retooling the operation equipment of 2–3 small workshops to be supported by the project.

Training on E&S issues: including dates, number of trainees, and topics.

Footprint management: details of any work outside the boundaries or major offsite impacts caused by ongoing construction, to include date, location, impacts, and actions taken.

External stakeholder engagement: highlights, including formal and informal meetings, and information disclosure and dissemination, to include a breakdown of women and men consulted and themes coming from various stakeholder groups, including vulnerable groups (e.g., disabled, elderly, children, etc.).

Details of any security risks: details of risks that the contractor may be exposed to while performing its work; the threats may come from third parties external to the project.

Worker grievances: details including occurrence date, grievance, and date submitted; actions taken and dates; resolution (if any) and date; and follow-up yet to be taken; grievances listed should include those received since the preceding report and those that were unresolved at the time of that report.

External stakeholder grievances: grievance and date submitted, action(s) taken and date(s), resolution (if any) and date, and follow-up yet to be taken; grievances listed should include those received since the preceding report and those that were unresolved at the time of that report. Grievance data should be gender-disaggregated.

8.12.19 Major changes to contractor's environmental and social practices

Deficiency and performance management: actions taken in response to previous notices of deficiency or observations regarding E&S performance and/or plans for actions to be taken; these should continue to be reported until the owner determines that the issue is resolved satisfactorily.

8.12.20 Environmental and social liabilities of the contractors

Contractors will be legally and financially accountable for any environmental or social damage or prejudice caused by their staff, and thus, they are expected to put in place controls and procedures to manage their environmental and social performance. A breakdown for the cost of noncompliance for each mitigation measure will be enclosed in the bidding documents. These will include:

- Mitigation measures to be included in the contract will be specified in the project ESMP.
- Deductions for environmental noncompliance will be added as a clause in the bill of quantities (BOQ) section.
- Environmental penalties shall be calculated and deducted in each submitted invoice.
- Any impact that is not properly mitigated will be the object of an environmental/social notice by the owner.
- For minor infringements and social complaints or an incident that causes temporary but reversible damage, the contractor will be given a notice to remedy the problem and restore the environment. No further actions will be taken if the project engineer confirms that restoration is done satisfactorily.
- For social notices, the project engineer will alert the contractor to remedy the social impact and to follow the issue until solved. If the contractor does not comply with the remediation request, work will be stopped and considered under no excused delay.
- If the contractor hasn't remedied the environmental impact during the allotted time, the project engineer will stop the work and give the contractor a notification indicating a financial penalty according to the noncomplied mitigation measure that was specified in the bidding document.
- No further actions will be required if the project engineer sees that restoration is done satisfactorily. Otherwise, if the contractor hasn't

remedied the situation within one day, any additional days of work stoppage will be considered no excused delay.

- Environmental notifications issued by the project engineer might include one or more environmental penalties.
- In the event of repeated noncompliance totaling 5% of the contract value, the project engineer will bring the environmental and social notices and the deduction history to the owner in order to take legal action.

8.12.21 Management of hazardous waste

A waste management plan should be prepared and implemented for safe management of waste. There are two essential prerequisites to ensure that the plan is implemented–sufficient staff and financial resources. The plan covers all waste life cycles, that is, generation, storage, collection, transportation, and finally safe treatment or disposal. It is anticipated that solid waste will comprise paper, wood, plastic, scrap metals, and glass. The overall volume is expected to be low.

Generation: Solid waste generated during construction mainly consists of municipal and construction wastes that shall be stored in a separate container. Waste should be separated from hazardous waste. In addition, as some waste can be reused, then waste segregation should be carried out. For example, wood waste will be given to factories that reshape it and make it usable for fireplaces for residential purposes.

Collection: Solid waste will be collected by a local joint service council using special vehicles.

Disposal: collected waste should be disposed of in an authorized landfill.

A waste management procedure will be revised to be aligned with the local solid waste management strategy.

8.13 Hazardous waste management plan for PVPS

Hazardous waste will likely be comprised of fuel, oils, lubricants, hydraulic/insulating fluids, batteries, tires, metal drums, empty chemical containers, and PV panels. Scrap metals, plastic, batteries, metal drums, old meters, and glass waste will be awarded to small factories that recycle them in a safe way.

According to international environmental law, the law forbids anyone from handling (manufacturing, storing, distributing, using, treating, disposing) hazardous materials or waste except according to the regulations and instructions in coordination with competent parties. Therefore, it is essential to have a hazardous waste management plan that consists of the following:

(a) Any hazardous waste generated as a result of any activity during construction or by the end of the project life should be stored in separate containers. The containers should be labeled "hazardous waste." The labeling system should be clear and well known to the public and workers to ensure general safety.

(b) Transportation of hazardous waste containers should be with special vehicles by special contractors. Before transporting this hazardous waste, a form should be filled out by the generator and transporter indicating the amount and type of hazardous waste. Written permission for transporting the hazardous waste to a registered treatment facility should be issued by the EQA.

(c) Transboundary of hazardous waste is not allowed unless written permission is issued by the EQA. The permission complies with Basel convention requirements.

(d) Hazardous waste record keeping should be created and checked by the ESO from time to time to make sure that hazardous waste is well managed.

(e) Disposal of the PV panels and other hazardous waste such as used oil will be according to EQA hazardous waste management regulations that meet Basel convention requirements.

(f) Existing technical facilities for treating and disposing of hazardous waste should be assigned before the start of the project.

(g) For emergency cases, all workers expected to be in contact with hazardous waste should be trained for safe handling of hazardous waste.

(h) All workers should be familiar with hazardous waste warning signs.

8.14 Grievance mechanism for workers

Contractors will put in place a grievance mechanism for workers that is proportionate to the workforce, according to the following principles:

Provision of information: All workers should be informed about the grievance mechanism at the time they are hired, and details about how it operates should be easily available, for example, included in worker documentation or on notice boards.

Transparency of the process: Workers must know to whom they can turn in the event of a grievance and the support and sources of advice that are available to them. All line and senior managers must be familiar with their organization's grievance procedure.

Keeping it up to date: The process should be regularly reviewed and kept up to date, for example, by referencing any new statutory guidelines as well as changes in contracts or representation.

Confidentiality: The process should ensure that a complaint is dealt with confidentially. While procedures may specify that complaints should first be made to the workers' line manager, there should also be the option of raising a grievance first with an alternative manager, for example, a human resources (personnel) manager.

Nonretribution: Procedures should guarantee that any worker raising a complaint will not be subject to any reprisal.

Reasonable time scales: Procedures should allow for time to investigate grievances fully but should aim for swift resolutions. The longer a grievance is allowed to continue, the harder it can be for both sides to get back to normal afterward. Time limits should be set for each stage of the process, for example, a maximum time between a grievance being raised and the setting up of a meeting to investigate it.

Right of appeal: A worker should have the right to appeal to the owner or national courts if he or she is not happy with the initial finding.

Right to be accompanied: In any meetings or hearings, the worker should have the right to be accompanied by a colleague, friend, or union representative.

Keeping records: Written records should be kept at all stages. The initial complaint should be in writing if possible, along with the response, notes of any meetings, the findings, and the reasons for the findings.

Relationship with collective agreements: Grievance procedures should be consistent with any collective agreements.

Relationship with regulation: Grievance processes should be compliant with the national employment code.

8.15 Responsibilities for environmental and social monitoring

The owner's PMU is responsible for compliance with any provisions of environmental and social documents required under the ESF. He/she will be updating the drafted documents as soon as the components are more specifically determined. The ESO will also be responsible for monitoring the plans and, along the course of project implementation, the need to assign ESOs.

The EQA will be fully informed and any additional requirements will be followed between the ESO and the EQA. The owner will be responsible for grievances reported from communities and workers related to the implementation of project activities.

8.16 Monitoring evaluation and reporting

The objectives for monitoring are:

- To record environmental impacts resulting from the subproject activities and to ensure implementation of the "mitigation measures" identified earlier in order to reduce adverse impacts and enhance positive impacts from project activities.
- To alert project authorities by providing timely information about the success or otherwise of the EIA process as outlined in this ESMF in such a manner that changes to the system can be made in a timely manner, if required.
- To make a final evaluation to determine whether the mitigation measures designed into the subprojects have been successful in such a way that the presubproject environmental and social conditions have been restored, improved upon, or are worse than before.

Environmental monitoring needs to be carried out during preconstruction, construction, and postconstruction of the subprojects in order to measure the success of the recommended mitigation measures.

The preconstruction stage should ensure that:

- Proposed construction activities applicable at each site are subjected to environmental screening, and that the plan and design for construction activities confirms to the environmental guidelines
- A site-specific environmental assessment (ESMP or ESIA) is prepared on time and incorporated into bidding documents.

Construction phase: ESO at the owner will conduct compliance monitoring, using the specific environmental measures relevant to and prescribed for the activities as well as to assess general environmental management/performance. A report should contain information with regard to environmental compliance in accordance to the provisions of the tender document as well as any difficult or outstanding works that need to be prepared. A monitoring plan should be prepared by the ESO.

Apart from general monitoring of mitigation/enhancement measures and health and safety protocols (as outlined in the ESMF and tender document), important environmental parameters to be monitored during the construction phase of the subprojects include noise level, drainage congestion, and traffic problems. However, the requirement and frequency of monitoring would depend on the type of subproject and field situation. For certain subprojects (e.g., rehabilitation of existing distribution line), monitoring of these parameters is not critical while monitoring of some of these parameters (e.g., noise level) would be needed only if significant pollution is suspected. Table 8.4 presents guidelines for monitoring of specific environmental parameters during the construction phase of different subprojects. In addition to Table 8.4, describe the routine monitoring work hat will be done by the owner's ESO to ensure that:

- All personnel at work sites shall be provided with protective gear such as helmets, goggles, boots, etc., so that injuries to personnel are avoided or minimized.
- As the workforce is likely to be exposed to noise levels beyond regulatory stipulated limits, they shall be provided with protective gear such as ear plugs, etc., and regularly rotated.
- Dust suppression measures such as sprinkling of water shall be ensured at all operation areas.
- The work and campsites shall have suitable facilities for handling any emergency situation such as fire, explosion, electrocution, etc.
- All areas intended for storage of hazardous materials shall be quarantined and provided with adequate facilities to combat emergency situations. All required permits for the storage of inflammable/hazardous materials are to be obtained.
- The construction workers, supervisors, and engineers shall be properly trained and with sufficient experience.
- The operational areas shall be access controlled and entry shall be allowed only under authorization.

Postconstruction phase: ESO at the owner will prepare a summary report for the implementation effectiveness of all environmental and social mitigation measures and share it with stakeholders and communities.

The following are some of the pertinent parameters and verifiable indicators that can be used to measure ESMP process, mitigation plans, and performance.

- Has the project resulted in better living standards for the community?
- How has the adoption of the ESMF requirements improved the environmental health and biophysical state of the people?
- Has the project resulted in job creation?

Table 8.4 Environmental and social indicators and monitored institutions.

Impact	Indicator	Information source	Responsibility
Air emissions	Reduction in GHG caused by the project's cumulative reduction in GHG emissions from project installations	Contractor	The owner
Noise emissions	Noise intensity and duration in installation of solar panels	Contractor	The owner, JDECO
Heat or light reflection	Number of complaints on heat or light reflection	Contractor	The owner, EQA
Biodiversity	Number of trees that have fallen Reported number of incidents of injury and killing of birds	Contractor	The owner, EQA
Chemicals	Number of recorded accidents due to chemical exposure	Contractor	The owner, EQA
Cultural heritage	Reported complaints on the reduction of aesthetic value or impact on heritage	Contractor	The owner, EQA
Employment	Number of technical and unskilled workers hired and contract duration		The owner
Economic and livelihood impacts	Individual project cost savings to utility company Price of solar panels purchased per kW Cumulative cost savings to government Changes in household incomes in project locations Proportion of household expenditure on electricity		The owner
Social conflicts	Number of stakeholders consulted and minutes of the meeting Number of complaints received on inconvenience and maintenance	Contractor	The owner, EQA
Health and safety	Quantity of day -to-day waste produced and taken to waste management site Quantity of solar panel special waste taken to designated waste Quantity of solar panel waste exported	Contractor	The owner, EQA

- Has ESMF adoption resulted in sustainable energy use and improved efficiency?
- Are periodic monitoring reports being completed?
- Are processes defined in the ESMF working well?
- How many complaints/grievances have been received regarding the project?

Table 8.4 shows some specific environmental and social indicators that need to be monitored and assessed by various institutions.

The implementation effectiveness will be carried out by the owner. This will be undertaken during midterm and the end of the project. The evaluation will assess the ESMF's effectiveness in addressing the environmental and social impacts of the project. The midterm evaluation will give feedback for implementation of the ESMF.

Monitoring frequency should be once a week, especially during heavy equipment tranfeer, or when drainage/traffic congestion is suspected.

8.17 Capacity building and training requirement

A qualified and trained staff at the owner is very essential to ensure that mitigation measures have been applied, implemented, and are functioning as intended. This means that staff from the above institutions has:

- Training and awareness sessions related to environmental management, policies, and regulations.
- Training on environmental assessment and monitoring activities specific to energy projects.
- Training sessions on preparing manuals and guidelines on how to assess the environmental impact of ASPIRE project components.
- Training on how to record and resolve any grievance.

8.18 ESMF implementation budget

The cost of mitigation measures of the ESMP, which is part of the project tender documents, will be included in the BOQ and the bidding documents.

Examples of mitigation measure costs include:

- Costs of dust suppression during excavation work and costs of monitoring noise during construction (shall be calculated based on the frequency of monitoring and cost of equipment);
- Costs of installing erosion control measures shall be estimated as part of the engineering costs;

- Cost of labor management and labor and community health and safety shall be estimated as part of the engineering costs;
- Cost of waste management procedures shall be estimated as part of the engineering costs;
- Cost of clean-up and disposal of construction debris and waste shall be estimated as part of the engineering costs;
- Cost of implementation of stakeholder engagement and grievance mechanism; and
- Costs related to hiring and training of ESO.

The owner will be responsible for financing the environment and safety officer post and monitoring activities as part of the project administrative costs. It may be useful to institute monitoring milestones and provide resources, as necessary, to carry out the monitoring activities. Also, the proposed indicators may be further elaborated and validated to accommodate any significant site-specific needs, in each case with the input and oversight of the EQA and in compliance with the World Bank ESF.

8.19 Social management framework

This section of the social management framework (SMF) will provide guidelines regarding the mechanism to identify and address concerns and impacts resulting from the project's activities during the project's life cycle. The SMF also aims to ensure that a transparent and inclusive approach is integrated into the project's activities by engaging all potential stakeholders, starting from the planning phase and throughout the project's life cycle, to ensure sustainable outcomes. The adoption of a transparent and inclusive approach ensures the early identification of potential social risks that might be caused by the project's activities and allows the stakeholders to participate in finding mitigation solutions to them. Broader participation and engagement of key stakeholders, public transparency, and institutional accountability foster a sense of ownership over the project, allowing early identification of social risks and contributing to designing comprehensive mitigation strategies to them.

8.19.1 Social management principles

8.19.1.1 Inclusion

The project should ensure that disadvantaged or vulnerable groups in the communities are identified and involved in the project activities, starting from the planning phase to postimplementation. The opinions, expectations, and concerns of these groups are as equally important as of the wider population in any community where the project's activities will take place. These include the very poor, women, minority communities, disabled people, and people with special social or cultural characteristics.

8.19.1.2 Participation

The communities should have the opportunity to have a deep and positive exchange of information and be fully engaged in the project's activities. Stakeholders should have full access to the project's information and have the opportunity to participate in designing the project's activities and monitor its progress as well as provide feedback when needed. Communication between project stakeholders and the project's implementing partners should be both ways rather than one way.

8.19.1.3 Transparency

Stakeholders should be able to have full access to information concerning the project's activities. The project implementing partners should disclose the project information through domains that are accessible and known to the stakeholders, taking into consideration their different educational levels. The disclosed information includes all project information, including but not limited to reports from the initial environmental and social screening visits and consultation meetings as well as a resettlement action plan, where applicable.

8.19.1.4 Social accountability

Social accountability will be an integral component in the project's design, allowing the local community to be fully engaged in the project. Strengthening transparency and accountability includes strengthening the grievance redress and management systems. Responses to grievances should be given in a timely manner.

8.19.1.5 Social safeguards

The following social safeguarding principles, among others, will be followed and incorporated in all mitigation plans consistent with the requirements of the RF and ESS5:

- All project affected parties (PAPs) will be compensated for losses resulting from project interventions.

- All compensation will be at replacement value; that is, the current market price at which the asset can be replaced, without deducting depreciation and salvage value.
- The project implementing partners will announce the cut-off date and will determine the people who will be included as PAPs. The public consultation and communication component will be an ongoing activity of the project.
- A clear grievance redress mechanism (GRM) will be developed and publically announced through portals known to all PAPs, allowing them to file their grievances and ensuring a fair, appropriate, and comprehensive solution to their problems.

8.19.2 Social management procedure

The social management plan of the proposed project will start with the identification of subprojects followed by social and environmental baseline screening of the subprojects. Based on the social and environmental screening, a social assessment plan would be determined. If a subproject is found to have no significant social safeguard issues, including loss of land, assets, or source of income, only a social safeguard report will be prepared summarizing the findings of the screening. However, if the screening identifies social safeguard issues, the subproject would require a social impact assessment along with the preparation of a resettlement action plan (RAP).

8.19.3 Grievance redress mechanism

A specific GRM should be set up for the project. The GRM is essential to allow individuals who believe the project's activities will have adverse effects on their livelihoods, assets, or wellbeing to have access to fair and just solutions to their concerns. The GRM would include the formation of a special committee that will be responsible to:

- Receive, record, and sort grievances;
- Conduct an initial assessment of the grievances;
- Refer grievances to appropriate units or people;
- Follow up with the filed grievances;
- Closure of grievances.

The committee's members should receive a capacity building training on how to receive, handle, respond, and close the filed grievances in line with best international practices.

The GRM should also include the production of written information material, including pamphlets and posters, which include information about the project as well as the GRM. Possible portals to submit grievances include suggestion boxes in local council headquarters, the local mosque, and social media pages such as Facebook. A designated database should be established to handle the filed grievances.

8.19.4 Resettlement framework

A resettlement framework (RF) is prepared as a separate document for the project. The RF sets out the policies, principles, institutional arrangements, schedules, and indicative budgets that will take care of anticipated resettlements for various project components. These arrangements ensure that there is a systematic process for ASPIRE's implementation that assures continuous beneficiary participation, involvement of relevant institutions and stakeholders, and adherence to national procedures and requirements as well as outline entitlement and compensation for affected persons. RF has attempted to identify the PAPs, the types of impacts, and strategies for compensation and/or restoration of potential losses for individuals and businesses. RF provides guidance to the owner to establish the mechanism to compensate losses adequately according to the correspondent legislation and ESS5 requirements, and to apply the project activities with the least disturbance to the communities hosting the project.

RF also applies to other activities resulting in involuntary resettlement that in the judgment of the bank are directly and significantly related to the bank-assisted project, necessary to achieve its objectives, and carried out or planned to be carried out contemporaneously with the project. The eligibility criteria will ensure that all PAPs are clearly recognized as eligible for assistance as per the provisions of the RF for the land that they occupy, or their livelihoods or assets that are affected. As required by the RF, consultation with local authorities and PAPs at various site locations is conducted by the owner. The participants are informed about the project's salient features and possible impacts to the local population. The feedback and areas of concerns provided by the participants is carefully recorded for further action.

The owner will submit the subproject RAP (where required) to the bank for review and clearance before implementation. The owner, upon approval from the bank, will implement the RAP with assistance from the consultants and the ESO, surveyors, and overseers. An individual payment plan will be prepared for each affected person and mitigation plans, including replacement of affected physical structures by the owner, will be also documented as a reference for future tracking. All declarations and agreements as per the RAP will be executed before taking over land through voluntary contribution, direct purchase, or exchange, and disclosed to the public.

For RAP implementation, the owner's ESO will be responsible to follow up and monitor that the RAP is implemented according to the timings associated with the project resettlement process. The owner will also be responsible for performance monitoring against previously set milestones, including the number of public consultation meetings held, the censuses of individuals, the assets and inventories conducted, and the number of housing units allocated.

The owner will also recruit an external local consultant to conduct an external monitoring of the resettlement process. The consultant will monitor whether the PAPs have been fairly compensated and their incomes and livelihoods reestablished to preproject levels or better.

8.19.5 A stakeholder engagement plan (SEP)

A stakeholder engagement plan (SEP) has been developed for the project that seeks to define a technically and culturally appropriate approach to consultation and disclosure. The goal of this SEP is to improve and facilitate decision-making and create an atmosphere of understanding that actively involves project-affected people and other stakeholders in a timely manner, and that these groups are provided sufficient opportunity to voice their opinions and concerns that may influence project decisions. The SEP is a useful tool for managing communications between the owner and stakeholders.

For the transmission line and connection point components, there will be a need to communicate directly with owners of lands who will be affected by clearance of the road reservation. It is not possible to identify these individuals at this stage, but the owner will be responsible for preparing a communication plan to discuss potential impacts and agree to the timing for transmission clearance activities when the exact location of the subprojects is identified.

The public consultation and stakeholder engagement (PC and SE) seek to define a technically and culturally appropriate approach to consultation, disclosure, and grievance redress. The goal of this stakeholder engagement (SE) is to improve and facilitate decision-making and create an atmosphere of understanding that actively involves project-affected people and other stakeholders in a timely manner, and that these groups are provided sufficient opportunity to voice their opinions and concerns that may influence project decisions. Therefore, the SE is a useful tool for managing communications between the owner and stakeholders.

The SEP will be updated once the exact locations of subprojects are identified and the ESO shall be engaged by the project to liaison with the local stakeholders and project-affected community.

Table 8.5 illustrates the needs and concerns raised by the community during the consultation activities conducted over different time scales. The owner should address these raised concerns.

8.19.6 Grievance redress mechanism

The GRM addresses grievances in an efficient, timely, and cost-effective manner that arise in the project, either due to actions by the owner or the contractor/subcontractors employed by the owner, from affected communities and external stakeholders. A separate mechanism is developed to address worker grievances. The owner is responsible for managing the GRM, but many of the grievances on the project will likely relate to the actions of the contractor and so will need to be resolved by the contractor. The owner with the support of the implementation consultant will administer the GRM process, deciding whether they or the contractor is responsible and determining the best course of action to resolve the grievance. The implementation consultant will support the owner to monitor grievance resolutions being undertaken by the contractor.

The project GRM deals with the issues of land and other asset acquisitions (e.g., amount of compensation, suitability of residual land plots, loss of access roads, loss of livelihood, etc.) as well as the losses and damages caused by construction works, and any direct or indirect environmental and social impacts. Therefore, the GRM has to be in place by the time the owner starts preparation of the RP (if applicable) and the ESIA, and shall function until the completion of all construction activities and beyond until the defect liability period ends. PAPs and other potential complainants should be fully informed of the GRM, its functions, procedures, timelines, and contact persons both verbally and through booklets and information brochures during consultation meetings and other stakeholder engagement activities. The owner will keep a log of complaints at hand.

Table 8.5 Analysis of concerns and mitigation measures.

Fear/Concern	Mitigation measures
Fear for public safety due to magnetic fields, malfunctions in lines	- The proposed medium voltage network lines have been designed horizontally so they are at a distance of no less than 2 m from buildings in accordance with international standards - The proposed medium voltage electrical networks do not pass over buildings and houses - Prepare a labor management procedure (LMP) and occupational health and safety (OHS) guidelines - Community health and safety measures - Raise public awareness on the contractor's responsibility commitment to the ESMPs and LMP where child labor bellow age 18 will not be allowed
Fear that the land would lose its value due to the project's	- Prepare a stakeholder engagement plan (SEP) - Plan land use change compensation, relocation - Prepare a resettlement action plan (RAP)
Fear of affecting the village flora	- In designing the project's activities, including the towers' proposed sites, the project shall take into consideration that it does not bring harm to vegetation, including old trees or green areas - Avoid installation sites that require cutting or substantially pruning a protected tree, an old tree, or a known bird-nesting tree - In some specific sites where trees are to be removed to build new connection points, the trees will have to be planted in the same type of soil in another place

Continued

Table 8.5 Analysis of concerns and mitigation measures—cont'd

Fear/Concern	Mitigation measures
Fear of losing access to private land or inability to build or construct on the land	- The most preferable temporal routes will be considered prior to construction time after authorization (use of existing routes will be the priority as far as possible) - A census and asset survey will be undertaken as part of the ESIA and RAP to identify the eligible PAPs and determine the magnitude and significance of the land impact - Local authorities of affected areas and land owners within the route will be approached by the project manager and the ESO for an explanation of the project scope and the benefits the PAP will gain - -All affected will be identified and compensated as per the resettlement policy framework and subsequent RAP - Stakeholder consultations will be undertaken in line with the SEP - A grievance mechanism will be implemented and monitored
Fear that the land would lose its value due to the project's activities,mainly the towers	- A survey team shall ensure that during the survey and tower pegging processes the tower to be established causes no damage to the habitat of existing communities - A marked route pegs should be designed to guide awarded contractor on appropriate route to take - A Grievance Mechanism will be implemented and monitored - Plan land use change compensation, relocation - Any affected land should be rehabilitated directly after constructing the towers

Table 8.5 Analysis of concerns and mitigation measures—cont'd

Fear/Concern	Mitigation measures
Fear that the project would endanger lives because they live in semipermanent homes made of corrugated metal	- The survey team shall ensure that during the survey and tower pegging processes, the tower to be established causes no damage to the habitat of existing communities - Prepare a labor management procedure (LMP) and occupational health and safety (OHS) guidelines - Community health and safety measures should be prepared and applied
Fear of increased charges on electricity service	- Electricity price should be regulated - Before setting electricity service prices, living standards should be taken into consideration
Lack of trust in responsible stakeholders	- The distribution companies as service providers should work to restore their credibility among the end users - In designing the project's activities, the company shall ensure that the village residents are engaged and well informed of the project's sites activities - Distribution companies should make sure that erected towers will be within the street's right of way - Distribution company should raise the public's knowledge about the project through public sessions, media, radio spots, etc. - Prepare a grievance mechanism and make sure it is clear to the public
Fear that the project's activities will trespass over the farmers' private land and not within the road's RoW	- Redesign the route path to make sure that the project's activities will not trespass over farmers' private land. If this cannot be avoided, then a time schedule for activities to be carried out should be prepared and proposed for private land to approve before work is started

Continued

Table 8.5 Analysis of concerns and mitigation measures—cont'd

Fear/Concern	Mitigation measures
	- Distribution companies should make sure that erected towers will be within the street's right of way - A census and asset survey will be undertaken as part of the ESIA and RAP to identify the eligible PAPs and determine the magnitude and significance of the land impact - Make sure that even when the existing road is widened to 20 m, as it was designed for, the location of new towers will not trespass on private land - A grievance mechanism for affected areas and land concerns within the route will be implemented and monitored by local authorities, the project manager, and the EOS
The project would contribute to economic activity in the area such as quarries that would disturb our way of living	- Other surrounding activities should comply with OSH guidelines in order not to affect farmers
Unfair employment practices.	- Priority for employment within the project activities should be given to local people but should comply with the LMP.

Typical grievances related to the project activities are:

- Land acquisition and physical displacement.
- Loss of land value due to project activities.
- Loss of access to private properties or assets due to project activities.
- Physical damage to health and well-being during project construction phase and postconstruction phase.
- Damage to residents' source of income such as crops, trees, or livestock.

The owner will implement an effective GRM with the objective of helping third parties to avoid resorting to the judicial system as far as possible. Complainants can seek redress from the judicial system at any time. The step-by-step process does not deter them from approaching the courts. All grievance-related correspondence will be documented and the grievance resolution process will be systematically tracked.

The ESO will be assigned to follow up on complaints related to the project. The complaint, in order to be filed, should be related to the project components and/or to its implementation and management. The grievance resolution process involves the following main steps:

- Anyone from the affected communities or anyone believing they are affected by the project can submit a grievance by completing a written grievance registration form that will be available.
- Where possible, it is desirable that complaints are submitted in writing by the complainant. Should the complainant not wish to comply with this request and submit the complaint verbally, then the complainant information and the details of the complaint should be entered in the GRM log.
- The complainant fills in the designated form in writing and signs it, or fills it out electronically, including all personal information and details of the complaint.
- The complainant encloses all copies of documents that may support the complaint.
- The GRM staff at the complaints unit will ensure that the form is filled in accurately. The complainant receives a receipt or a confirmation email of acknowledgment with a reference number to track the complaint.

If the complainant chooses to file his/her complaint verbally, the GRM employee must register the complainant information and details of the complaint into the system. The complainant will receive a reference number to track his/her complaint.

The GRM staff will enter the complaint into the GRM log. The complaints register records the following information:

- Complaint reference number.
- Date of receipt of complaint.
- Name of complainant.
- Confirmation that a complaint is acknowledged.
- Brief description of complaint.
- Details of internal and external communication.
- Action taken, including remedies /determinations/result.
- Date of finalization of complaint.

Original documentation must be kept on file.

A GRM system will be established that includes a GRM committee. The ESO will inform the complainant that an investigation is underway within three business days. The complainant shall be informed of the estimated duration for resolving the complaint, which is no later than

10 business days from the date of receipt of the complaint. Where the complaint is unlikely to be resolved within the estimated duration, the ESO must promptly contact the complainant to request additional time and explain the delay. In any event, the complaint must be resolved no later than 2 weeks from the date of receipt of the complaint. If the complaint is not, the ESO will refer the complaint to the director of the PMU to take the appropriate measures.

The ESO will then follow the steps below:

- Verify the validity of the information and documents enclosed.
- Ask the complainant to provide further information if necessary.
- Refer the complaint to the relevant department.
- The ESO in charge in the relevant department shall conduct field visits for verification, if necessary, and prepare recommendation to the PMU director of actions to be taken and of any corrective measures to avoid possible reoccurrence.
- The ESO shall register the decision and actions taken in the GRM log.
- Notifying the complainant:

The ESO shall notify the complainant of the decision/solution/action immediately, either in writing, by calling, or by sending the complainant a text message.

When providing a response to the complainant, the ESO must include the following information:

- A summary of issues raised in the initial complaint;
- Reason for the decision.
- Closing the complaint:

A complaint is closed in the following cases:

- Where the decision/solution of the complaint is accepted by the complainant, the ESO shall close the complaint and sign the outcome and date in the complaint register.
- A complaint that is not related to the project or any of its components.
- A complaint that is being heard by the judiciary.
- A malicious complaint.

Where the complainant is not satisfied with the outcome of his/her complaint, the following procedures shall be considered:

-Internal dispute resolution scheme

The ESO shall advise the complainants that if they are not satisfied with the outcome of their complaint, they may readdress the issues with the DG of the PMU at the owner and request a further review or consideration.

Where the complainants are not satisfied with the resolution provided by the DG of PMU, the ESO shall advise the complainants to readdress the issue either to the president or the owner.

-External dispute resolution scheme

In case the complainants are not satisfied with the internal procedures for handling complaints, the outcomes of the complaints, or for any unhandled complaints, the ESO shall provide information on a complainant's right to refer their complaint to the cabinet unit for grievances or to the judicial system.

The ESO shall review the complaints register regularly for the purpose of providing analysis and reports on complaints to the director of the PMU. The report shall include the number of complaints received, handled, and closed. It shall also include analysis on systemic and recurring problems. This will assist the project management in determining the cause of complaints and whether remedial action is warranted.

Periodic reporting shall be as follows:

- A monthly report to the project management at the PMCU.
- A quarterly or semiannual report to the project management.
- There is a complaints unit at the owner. The complaints unit headed by the ESO is regulated by the Council of Ministers Decision No. (8) of 2016 and by the Procedure Manual No. (20/17) of 2017. Both documents are made public and published in Arabic on the ministries' websites. A detailed GRM manual that includes guidelines on filing and handling complaints at the project level has been prepared. The owner will keep a log for grievances and how complaints were resolved within a stipulated time frame, and then produce monthly reports for senior management. Grievances/feedback reports include data on the number of grievances/feedback received, compliance with business standards, issues raised in grievances/feedback, trends in grievances/feedback over time, the causes of grievances/feedback, whether remedial action was warranted, and what redress was provided.

The owner will require contractors to develop and implement a grievance mechanism for their workforce prior to the start of civil works. The construction contractors will prepare their labor management procedure before the start of civil works, which will also include a detailed description of the workers grievance mechanism.

The workers grievance mechanism will include:

- A procedure to receive grievances such as a comment/complaint form, suggestion boxes, email, or a telephone hotline;

- Stipulated time frames to respond to grievances;
- A register to record and track the timely resolution of grievances;
- An assigned staff to receive, record, and track the resolution of grievances.

The workers grievance mechanism will be described in staff induction trainings, which will be provided to all project workers. Information about the existence of the grievance mechanism will be readily available to all project workers (direct and contracted) through notice boards, the presence of suggestion/complaint boxes, and other means as needed. The ESO will monitor the contractors' recording and resolution of grievances, and report these to the owner in the monthly progress reports.

8.19.7 Grievance redress mechanism (GRM)

In compliance with international standards and in order to ensure that disadvantaged or vulnerable needs are taken into consideration, and that they are reached, the owner will adopt several mechanisms such as publishing all information about the project in the local language, holding workshops or meetings at suitable locations that women can easily access, provide needed facilities in public meetings for the handicapped or people with disabilities, and conduct visits to Bedouin families. In addition, when designing the grievance mechanism, the owner should take into account the availability of needed recourse for vulnerable groups to give feedback or send a complaint; for example, if an Internet option is not available to women at villages, the project management units at the owner will provide them with alternative options such as a telephone number for the GRM. It will be publically announced through portals known to all PAPs, allowing them to file their grievances and ensuring a fair, appropriate, and comprehensive solution to their problems.

In addition, the owner should be responsible for any grievance reported from communities and workers related to the environmental and social issues from the implementation of project activities. The GRM will also be used for complaints in relation to resettlement aspects as well.

8.20 Labor management

Wages: The minimum wage limit is regulated according to local law. The specific minimum wages for workers in general should not be lower than 30$/day and 500 $/month for employees. A labor agreement will determine the form and amount of remuneration. Remuneration will be paid at least once a month.

The insurance made by contractors for the contracted workers will pay compensation to the contracted workers for work-related damage that causes any deterioration to the employee's health and will cover the subsequent necessary treatment costs.

Deductions from payment of wages will only be made as allowed by the national law, and project workers will be informed of the conditions under which such deductions will be made.

Working hours: The maximum number of hours per day that contracted workers must perform on the project is 8 h (Saturday through Thursday), and the allowed work week of 48 h. For direct workers, the number of hours per day is 7 h (Sunday through Thursday), and the work week is 35 h. The daily working hours shall be reduced by at least 1 h in all hazardous or health damaging occupations in addition to nightly jobs.

Rest breaks: The employees will have a 1 h meal break each workday, taking into consideration that the worker shall not work for more than 5 consecutive hours. The duration of rest between working days is 1 day on Friday for contracted workers and 2 days for direct workers.

Overtime work: The extra working hours should not exceed 12 h a week. The worker shall be paid the wage of 1.5 h for each extra working hour he/she works.

Leave: An employee will have the right to enjoy paid leave for at least 21 working days, sick leave of 14 days, and unpaid leave for 14 calendar days per annum. Leave does not include maternity leave, which is 70 days. The worker shall have the right to a paid leave on religious and official holidays, which is not considered or counted as annual leaves.

Women: Labor law includes provisions for the prohibition of discrimination between men and women. Employment of women is prohibited in the following jobs or under the following conditions: dangerous or hard work, extra working hours during pregnancy and during the first 6 months after delivery, and during night hours except for the work defined by the Council of Ministers.

Labor disputes: Labor law includes provisions for worker exemption from legal fees arising from work-related disputes and allows workers to unionize. A bipartite committee will settle any disputes that may arise from the implementation of the agreement. The court has jurisdiction over labor-related disputes.

8.20.1 Overview of labor legislation: Occupational health and safety

- All potential risks to project workers' health and safety will be identified by all parties who employs workers and develop and implement procedures to establish and maintain a safe working environment, including workplaces, machinery, equipment, and processes under their control;
- Contractors will prepare a detailed OHS plan for their correspondent contracts, including risk assessment, mitigation measures, method statements, and a system of monitoring and reporting.
- Appropriate protective measures will be provided. These measures include providing adequate personal protective equipment (PPE) at no cost to the project workers;
- Contractors will assign a health and safety officer at construction sites;
- Project workers will receive OHS training at the beginning of their employment. Training will cover the relevant aspects of OHS associated with daily work, including the ability to stop work without imminent danger and respond to emergency situations. Training records will be kept on file. These records will include a description of the training, the number of hours of training provided, training attendance records, and results of evaluations;
- The contractor will develop and implement a reporting system for any accidents, diseases, incidents, or near misses. Every incident will be reported to the contractor and investigated, and relevant measures will be designed to avoid the incident in the future. Also, remedies for adverse impacts such as occupational injuries, disabilities, and diseases will be provided.

The ESO to be hired at the owner to oversee the implementation of the environmental and social measures across the project will be responsible for the following:

- Implement this LMP to direct workers;
- Ensure that contractor(s) responsible for the construction work prepare their labor management procedure in compliance with this labor management procedure, and an OHS plan before the initiation of construction work;
- Monitor that the contractor(s) are meeting obligations toward contracted and subcontracted workers;
- Monitor the implementation of contractors' labor management procedures;

- Monitor that OHS standards are met at workplaces in line with the OHS plan;
- Monitor training of the project workers;
- Ensure that the grievance mechanism for the project workers is established and monitor its implementation;
- Monitor implementation of the workers code of conduct.

The contractors will be responsible for the following:

- Appoint a qualified environmental and social expert to prepare and implement project-specific labor management procedure and OHS plans as well as to manage subcontractor performance;
- Develop a labor management procedure and OHS plan that will apply to contracted and subcontracted workers. These procedures and plans will be submitted to ESO at the owner or the supervision engineering office for review and approval before the contractors mobilize for the construction phase;
- Contractors will supervise their subcontractors' implementation of labor management procedures and the OHS plan;
- Maintain records of the recruitment and employment process of contracted workers.
- Engage and manage the project workers;
- Communicate clearly the job description and employment conditions to contracted workers.
- Develop and implement a worker grievance mechanism and address the grievance received from the contracted and subcontracted workers;
- Have a system for regular review and reporting on labor and OHS performance.
- Deliver regular induction on HSE training to employees;
- Ensure that all contractor and subcontractor workers understand and sign the code of conduct prior to the initiation of work;
- After the bidding process is completed and the contractors are known, the LMP can be updated to include additional details about contractors as necessary.

The supervision will be carried out by the ESO or by an external supervision engineering firm to oversee labor and safety performance daily on behalf of the owner. The ESMP requires the supervision to employ a qualified environmental and social expert for such oversight and to report on environmental and social due diligence to the owner on a weekly basis.

8.20.2 Policies and procedures

The contractors will prepare labor management procedures in line with this labor management procedure based on labor law. The principles and procedures presented below represent the minimum requirements, but are not an exhaustive list of requirements. The employment of project workers will be based on the principles of nondiscrimination and equal opportunity. There will be no discrimination with respect to any aspects of the employment relationship, such as recruitment, compensation, working conditions, terms of employment, access to training, promotion, or termination of employment. The following measures will be developed by the contractors and monitored by the owner or the supervision engineering firm that will be hired on behalf of the owner to ensure fair treatment of all employees:

- Recruitment procedures will be transparent, public, and nondiscriminatory with respect to ethnicity, religion, sexual orientation, disability, and gender;
- Clear job descriptions will be provided in advance of recruitment and will explain the skills required for each post;
- All workers will have written contracts describing the terms and conditions of work and will have the contents explained to them. Workers will sign the employment contract. Terms and conditions of employment will be available at work sites;
- Employees will be informed at least 2 months before their expected release date of the coming termination;
- The contracted workers will not pay any hiring fees. If any hiring fees are to be incurred, these will be paid by the contractor;
- The contracts will be developed in the Arabic language;
- In addition to written documentation, an oral explanation of conditions and terms of employment will be provided to workers who may have difficulties with understanding the documentation;
- The owner will include in contracts that all contractor (and subcontractor) personnel must be 18 years or older.

The main health and safety risks will be encountered by the construction contractors' workforce. Contractors bidding for the work will have to demonstrate the capability to manage health and safety risks and provide corresponding documentation. After the contract is awarded, the contractors are required to provide the labor management procedures and occupational health and safety plan in line with the ESMP. The contractors will ensure that occupational health and safety plans are implemented by subcontractors.

The owner will include into the bidding documents specific OHS standard requirements that all contractors and subcontractors will meet under the project. The standards will be consistent with local regulations, WBG EHS guidelines, and GIIP (good international and industry practices). The following OHS standard requirements should as a minimum be included in the OHS plan to be prepared by the contractors:

- Provide a safe workplace and risk assessment procedure will be completed before the commence of any construction activities, and safety measures will be implemented in accordance with applicable safety standards;
- Emergency response procedure;
- Fall prevention and working at heights;
- Excavation safety, ladder safety; welding and cutting safety; crane and machinery safety; hand tool safety;
- OHS training;
- Present an OHS accountability matrix for all staff including the project manager, contract manager, OHS staff, foremen, and all employees with clear roles and OHS responsibilities;
- All contractors must have their own OHS staff that will be responsible for the implementation and supervision of the OHS program;
- PPEs and other preventive measures will be provided at no cost for employees;
- Biweekly OHS meetings will be conducted to discuss preventive measures, deviations, noncompliances, accidents, and corrective actions;
- Contractors will conduct internal OHS surveys and audits to verify compliance of OHS practices. Noncompliances will be documented and reported internally. A time frame for a corrective action will be set and followed up. Contractors will document and report to the supervision consulting office all accidents and illnesses with a day lost or more as well as fatalities or serious injuries that may happen at a work site;
- There must be onsite resources for first aid and for more serious injuries. There must be a preapproved health facility for medical treatment as well as appropriate transportation of injured workers;
- Contractors will control access to the construction site only to authorized people. Workers must be trained to perform hazardous work such as working at heights, confined spaces, welding, etc. All workers must complete at minimum an OHS induction to have access to the construction site.

The owner will include in the contract(s) as requirement for contractors to report on issues such as the number of accidents rates, severity rates, number of recurring noncompliances, fatalities and serious injuries, and penalties for noncompletion.

The owner or supervision engineering firm (on behalf of the owner) will conduct periodic supervision of contractor's OHS performance, including site visits, continuously. These supervisions will cover compliance with the above-mentioned standards, accidents, recommendations, and the progress of ongoing corrective actions.

The owner or supervision engineering firm will review and approve contractors' safety plans and procedures. The owner will inform the association promptly about any incident or accident related to the project that has or is likely to have a significant adverse effect on the environment, the affected communities, the public, or workers (labor, health and safety, or security incident, accident or circumstance) as soon as reasonably practicable for regular incidents. For serious incidents such as strikes or other labor protests, serious worker injuries or fatalities, project-caused injuries to community members, or property damage, the reporting should be done no later than 24h. The owner will prepare a report on the event and the corrective action and submit to the association within 30 calendar days of the event.

The construction contractor will develop and implement a code of conduct. He should also submit the code of conduct to supervision responsible parties for review and approval. The code of conduct will reflect the company's core values and overall working culture. The contractors will be required to provide a monthly report on the performance of labor and occupational health and safety issues, which will be reviewed by the supervision responsible parties.

The contractor should abide to the policies and procedures stated in Section 8.20.2 related to nondiscrimination with respect to any aspects of the employment relationship, such as recruitment, compensation, working conditions, terms of employment, access to training, promotion, or termination of employment. The contractor will prepare, adopt, and implement a standalone security personnel management plan consistent with the requirements of ESS4 in a manner acceptable to the association.

8.20.3 Age of employment

A child under the age of 18 will not be employed or engaged in connection with the project.

The contractors will be required to verify and identify the age of all workers. This will require workers to provide official documentation, which could include a birth certificate, IDs for those above 16, or birth certificates, medical or school records, or parent IDs for those of age 15–16. The contractor shall keep the records/documents, which will be checked onsite by the supervision engineering offices and the district engineers.

If underage workers are found working on the project, measures will be taken to immediately terminate the employment or engagement of the child in a responsible manner. A regular review and checkup will be conducted by the parties responsible for supervision to make sure no underage workers are still working on the project.

8.20.4 Terms and conditions

The terms and conditions applying to direct workers should be set so as to provide for the rights of employees who will be assigned to work on the project. The terms and conditions of part-time direct workers are determined by their individual contracts. The contractor should provide the workers with the following information:

- Information to workers regarding their terms and conditions of employment including hours of work, wages, overtime, compensation, benefits, holidays, leaves, etc.;
- Provide workers with adequate periods of rest per week as well as annual holiday and sick leave, as required by national law;
- Ensure nondiscrimination and equal opportunity in the project;
- Set out measures to prevent GBV and SEA in accordance with the ESF;
- Ban the use or support of child, forced, or compulsory labor;
- Workers should have signed contracts with clear terms as per the labor law.

The contractor's labor management procedure will be prepared by following this LMP and setting out terms and conditions for the contracted and subcontracted workers.

8.20.5 Grievance mechanism

The contractors shall be provided, within the bidding documents, clear grievance mechanisms for the workers who will be employed or engaged in connection with the project. The workers grievance mechanism will include: (i) a procedure to receive grievances such as a comment/complaint form, suggestion boxes, email, and telephone, (ii) stipulated time frames to respond to grievances, (iii) a register to record and track the timely resolution of grievances, and (iv) a responsible office/department to receive, record, and track the resolution of grievances.

The contracted workers shall be informed within the training to be provided to them by the contractor about the grievance mechanism, and how and to whom they can address their complaints in case they didn't get responses on their complaints from the contractors. Moreover, at work sites, there will be clear and understandable grievance procedures and mechanisms written on boards for the project. Grievance feedback shall be communicated with the complainant by telephone, fax, email, or in writing.

8.20.6 Contractor management

The owner will use the local procurement procedures and the Bank's 2017 Standard Procurement Documents for solicitations and contracts, which include labor and OHS requirements. The owner,after receiving bids from the contractors, ensures that the contractors are legitimate and licensed.

The environmental and social instruments, including ESMF, ESIA, ESMP, SEP, and this LMP form, are an integral part of the bidding documents to be issued to contractors and shall also be part of the awarded contracts to these contractors. In addition, proper training and orientation to contractors will be made by the owner at different stages of awarding and implementation of the project to ensure full understanding and compliance.

During the process of selecting contractors who will engage contracted workers, the owner may review the following information:

- Business licenses, registrations, permits, and approvals;
- Records of safety and health violations, and responses;
- Documents relating to a labor management system, including OHS issues;
- Worker certifications/permits/training to perform required work;
- Accident and fatality records and notifications to authorities;
- Proof of worker experience and enrollment in related projects;
- Worker payroll records, including hours worked and pay received;
- Enrollment of safety members and records of meetings; and
- Copies of previous contracts, showing the owner inclusion of provisions and terms.

The performance of contractors in relation to contracted workers, focusing on compliance with their contractual agreements (obligations, representations, and warranties) will be managed and monitored by the owner, besides the responsible parties and their assigned resident engineers. Regular supervision checkups will be conducted to ensure environmental and social compliance with the environmental and social instruments and labor management records and reports compiled by contractors. Contractor labor management records and reports may include: (a) a representative sample of employment contracts; (b) records relating to grievances received and their resolution; (c) reports relating to safety inspections, including fatalities and incidents and implementation of corrective actions; (d) records relating to incidents of noncompliance with national law; and (e) records of training provided for contracted workers to explain labor and working conditions and OHS guidelines and procedures.

8.20.7 Community workers

As mentioned in Section 8.20.2, the project may not engage community workers. Community workers are not currently used by the owner in any projects due to the specialized labor needs required.

8.20.8 Primary supply workers

The primary suppliers of the project materials, including cables, electromechanical equipment such as circuit breakers and switch isolators, wires, protection and control equipment, power poles, steel products, solar systems, construction materials, etc., shall be local companies and formal businesses buying materials that are subject to high standards from Israel and international companies. These sectors are not known to involve significant risks of child labor and forced labor. In all cases where primary suppliers will be engaged, contractors will be required to inquire during their procurement process whether the supplier has been accused or sanctioned for any of these issues and also their corporate requirements related to child labor, forced labor, and safety.

If there are any risks identified related to child and forced labor and safety, the owner will prepare the procedures to address these risks. Contractors will be vetted using a different form that screens the supplier in regard to compliance with taxes, certification, licensing, public liability certificates, and workmen compensation. A separate form requires that the primary supplier identify the company's permanent staff and declare any current or prior arbitrations as well as any criminal convictions. Suppliers will be subject to regular review.

Index

Note: Page numbers followed by *f* indicate figures and *t* indicate tables.

E

T

W

FSC
www.fsc.org
MIX
Papier | Fördert gute Waldnutzung
FSC® C083411

Kolibri 360 GmbH
Ferdinand-Jühlke-Straße 7
99095 Erfurt, Deutschland
produktsicherheit@kolibri360.de